Synthese Library

Studies in Epistemology, Logic, Methodology, and Philosophy of Science

Volume 519

The aim of *Synthese Library* is to provide a forum for the best current work in the methodology and philosophy of science and in epistemology, all broadly understood. A wide variety of different approaches have traditionally been represented in the Library, and every effort is made to maintain this variety, not for its own sake, but because we believe that there are many fruitful and illuminating approaches to the philosophy of science and related disciplines.

Special attention is paid to methodological studies which illustrate the interplay of empirical and philosophical viewpoints and to contributions to the formal (logical, set-theoretical, mathematical, information-theoretical, decision-theoretical, etc.) methodology of empirical sciences. Likewise, the applications of logical methods to epistemology as well as philosophically and methodologically relevant studies in logic are strongly encouraged. The emphasis on logic will be tempered by interest in the psychological, historical, and sociological aspects of science. In addition to monographs *Synthese Library* publishes thematically unified anthologies and edited volumes with a well-defined topical focus inside the aim and scope of the book series. The contributions in the volumes are expected to be focused and structurally organized in accordance with the central theme(s), and should be tied together by an extensive editorial introduction or set of introductions if the volume is divided into parts. An extensive bibliography and index are mandatory.

Damian Fernandez-Beanato

A Working Scientific Demarcation

Springer

Damian Fernandez-Beanato
Bristol, UK

ISSN 0166-6991 ISSN 2542-8292 (electronic)
Synthese Library
ISBN 978-3-031-99004-5 ISBN 978-3-031-99005-2 (eBook)
https://doi.org/10.1007/978-3-031-99005-2

This Springer imprint is published by the registered company Springer Nature Switzerland AG
The registered company address is: Gewerbestrasse 11, 6330 Cham, Switzerland

Author's Note on the Text

Parts of this book were previously published as academic articles in 2020 and 2021, and then in modified versions in my doctoral dissertation at the University of Bristol in 2022. Other parts of this book were included in that dissertation without having been published in articles before. In all cases, this book should be deemed to contain the corrected, full, latest, and current versions of those works as of October of 2025.

Preface

The problem of demarcating science from non-science is generally regarded as being unsolved. This monograph executes an analytical process of elimination of different demarcation proposals put forward since the professionalisation of the philosophy of science, explaining why each of those proposals is unsatisfactory or incomplete. Then, it executes an alternative multicriterial scientific demarcation project proposed by Mahner (2007, pp. 521–22; 2013, pp. 29–43). This project allows for the demarcation not only of science from pseudoscience, but also from other non-scientific fields. In this way, this work accomplishes an intensional scientific demarcation, for the first time in the history of the philosophy of science. This demarcation is a directly applicable practical decision procedure and tool to assess the scientificity or non-scientificity of disciplines, theories, cognitive fields, practices, attitudes, testimonies, etc., able in principle to be utilised and applied in any context, for example by governments, regulatory agencies, courts, tribunals, and health care and education providers.

Bristol, UK Damian Fernandez-Beanato

References

Mahner, M. (2007). Demarcating science from non-science. In T. Kuipers (Ed.), *Handbook of the philosophy of science: General philosophy of science – Focal issues* (pp. 515–575). Elsevier.

Mahner, M. (2013). Science and pseudoscience. How to demarcate after the (alleged) demise of the demarcation problem. In M. Pigliucci & M. Boudry (Eds.), *Philosophy of pseudoscience: Reconsidering the demarcation problem* (pp. 29–44). The University of Chicago Press.

Acknowledgements

This book is the result of over a decade of work. I thank, from the bottom of my heart, all those who taught me philosophy and history of science, as well as logics, and philosophy of language. At the University of Buenos Aires, I had the life-changing privilege of learning from (and, in some cases, working under) wonderful, top-tier teachers such as Rodolfo Gaeta, Nélida Gentile, Alejandro Cassini, Marcelo Leonardo Levinas, Eleonora Orlando, Eduardo Barrio, Santiago Ginnobili, Sergio Barberis, Bruno Borge, and Roberto Azar. At the University of Tres de Febrero (UNTreF), I was equally fortunate to learn from Christián Carman, Pablo Lorenzano, Federico Penelas, Eleonora Cresto, and Diego Ríos, among others.

James Ladyman encouraged me to pursue this investigation and accepted serving as its supervisor during its PhD dissertation stage, guiding me during research and writing with important suggestions that determined different aspects of my final approach. Samir Okasha, Johannes Stern, and several anonymous reviewers provided valuable advice and comments; and Martin Mahner, Maarten Boudry, and the late Mario Bunge graciously gave me extremely useful consultations. Nancy Cartwright came from Durham and was a wonderful examiner.

To all of the above I owe an intellectual debt. Any shortcomings in this work are my own.

Further thanks go to my students at the University of Buenos Aires and the University of Bristol who made teaching enjoyable for me, raised interesting points, and expressed personal appreciation.

This work has been partially funded by the University of Bristol.

I would also like to express my love for my wife, Cecilia, who supported me emotionally, intellectually, financially, and probably in even other ways during the elaboration of the present text. To her and to my other beloved family members—my parents, Diana and Alfredo; my sister, Erica; my niece, Charlotte; and my aunt Beba, who fed me corn-starch porridge—go my thoughts.

Competing Interests The author has no competing interests to declare that are relevant to the content of this manuscript.

Contents

About the Author

Damian Fernandez-Beanato is an epistemologically oriented professional philosopher and historian of science. He has earned multiple degrees, including a Licentiate degree in History from the University of Buenos Aires, a Master's degree in Epistemology and History of Science from UNTreF University, and a PhD in Philosophy from the University of Bristol, England. He has also published several peer-reviewed academic articles on the topic of this book.

He specialises in the general philosophy of science, particularly the method of science and the demarcation of science from non-science, which he has been researching full-time for over a decade. During his doctoral research at Bristol, he developed the first accurate method for identifying non-scientific disciplines, claims, and attitudes. This revolutionary technique is finally presented in full form in this book.

He has been teaching logics, methodology of science, general philosophy of science, epistemology, and scientific demarcation at different universities, on different continents, and for a wide range of audiences (students of every academic discipline, and from well-to-do students to inmates, and everything in between) since 2014.

List of Abbreviations

AD	Anno Domini
ADHD	Attention deficit hyperactivity disorder
BCE	Before the Common Era
CAB	Current academic biology
CAP	Current academic psychology
CAS	Current academic sociology
CF	Cognitive field
CM	Cognitive meaningfulness
CS	Common sense
DNA	Deoxyribonucleic acid
ESP	Extrasensory perception
EWSD	Epistemic-warrant scientific demarcation
GA	Genethliac astrology
IUPAC	International Union of Pure and Applied Chemistry
MN	Methodological naturalism
NASA	National Aeronautics and Space Administration
ON	Ontological naturalism
RNA	Ribonucleic acid
SBS	Science in the broad sense
SD	Scientific demarcation
USSR	Union of Soviet Socialist Republics
WMAP	Wilkinson Microwave Anisotropy Probe

Part I
Retackling the Demarcation Problem

Chapter 1
The Multicriterial Approach to the Problem of Demarcation

Abstract This chapter executes an analytical process of elimination by considering the historical development of the scientific demarcation problems, evaluating all the main previous scientific demarcation proposals, and explaining why each of those proposals is unsatisfactory or incomplete. It then presents a comprehensive scientific demarcation project in which, among other contributions, the proposals incipiently put forward by Mahner (2013) and Hansson (2013) are theoretically elaborated on, expanded, developed, and adjusted up to the point at which they are ready for execution.

Keywords Science · Scientific demarcation · Non-science · Pseudoscience · Scientificity

1.1 Introduction

The problem of how to demarcate science from non-science is generally regarded as being unsolved, although it has been deemed one of the most important extant problems in the philosophy of science. This work accomplishes a successful intensional[1] epistemic[2] demarcation of science, for the first time in the history of the discipline.

[1] "An intensional definition of 'science' would involve properties of [science or of] scientificity, as opposed to an extensional definition" (Fernandez-Beanato, 2020a, p. 97), which would be given by a list of all scientific fields, practices, attitudes, modes of procedure, etc.

[2] An epistemic demarcation of science is a demarcation between science and non-science with respect to what is "special about the knowledge claims and the modes of inquiry of the sciences", i.e., about exclusive or typical properties that these claims and methods enjoy "in a way that exhibits an epistemic guarantee or evidential basis that is [epistemically] safer for science than for non-science" (Laudan, 1983, p. 118). That science possesses this characteristic is fairly non-controversial, and as such is accepted in this work without defence (although there are exceptions, e.g., introspec-

This chapter was previously published in shortened form as Fernandez-Beanato (2020b) and (2020c). One paragraph is from Fernandez-Beanato (2021).

D. Fernandez-Beanato, *A Working Scientific Demarcation*, Synthese Library 519, https://doi.org/10.1007/978-3-031-99005-2_1

This demarcation is a directly applicable practical decision procedure and tool to assess the scientificity or non-scientificity of disciplines, theories, cognitive fields, practices, attitudes, testimonies, etc., able in principle to be utilised and applied in any context, for example by governments, regulatory agencies, courts, tribunals, and health care and education providers.

To sharpen and improve our understanding of science would be beneficial for human well-being. Scientific demarcation (SD) criteria would better inform reliable philosophical and scientific decisions and judgments pertaining to a wide array of everyday human activities, including educational, financial, medical, legal, regulatory, and political endeavours, so this project is of profound social relevance.

This work takes the main problems of SD to be those of establishing criteria based on reflection on science and its history to: i) most generally, intensionally demarcate science from non-science (i.e., the problem of defining "science"); and: ii) in particular, to intensionally demarcate science from pseudoscience.[3] These remain unsolved, but this chapter outlines a project to solve i), therefore providing the basis to solve ii) (see Note 3).

Sections 1.2–1.3.3 of this monograph execute an analytical process of elimination by considering the historical development of the SD problems, evaluating different SD proposals put forward since the professionalisation of the philosophy of science, and explaining why each of those proposals is unsatisfactory or incomplete. Some of the objections that I make are well-known, but a review of this literature is essential to address the problems at hand. The main aim of this chapter is to present a comprehensive SD project in which, among other contributions, the proposals incipiently put forward by Mahner (2013) and Hansson (2013) are theoretically elaborated on, expanded, developed, and adjusted up to the point at which they are ready for execution. Section 1.3.4 takes up, elucidates, and expands the alternative multicriterial SD proposal put forward by Mahner (2007, pp. 521–22, 2013, pp. 29–43). Section 1.4 criticises Hansson's work, introduces the concept of epistemic-warrant SDs (EWSDs), refutes the idea that EWSDs and diagnostics of pseudoscientificity are necessarily normative, indicates a territorial aspect of EWSDs, argues in favour of accepting two types of SDs (namely, EWSDs and territorial SDs), presents arguments in favour of a territorially broad SD, advances a

---

tive knowledge: a person can know her own conscious thoughts and feelings more easily, quickly, and accurately than science currently could).

[3] In line with common philosophical thought (Hansson, 2017), this work defines a "pseudoscience" as a non-scientific doctrine or field that is presented by its theoreticians or practitioners as if it were scientific or a science, or that is easily taken to be so by a given epistemic community (this last clause is needed to cover those cases in which a unit of analysis or agent should be treated as pseudoscientific even if that unit or agent is not literally presented as being scientific or does not literally claim to be scientific, such as for example pseudoarchaeologists who do not claim to be archaeologists; see Fernandez-Beanato (2023): "[Graham] Hancock's hypothesis [...] is still pseudoscience and pseudoarchaeology even though he [says that he] doesn't claim to be a scientist or an archaeologist because millions of people will mistakenly believe [that] he is right on [...] archaeological matters." (pp. 49–50)). My intended intensional demarcation of pseudoscientificity is not exhausted by this definition, in part because it depends on the intensional demarcation of scientificity, and for reasons detailed in Chap. 20.

definition of "science in the broad sense" (SBS, also known as broad science[4]), and indicates how properties of scientificity[5] of different levels are articulated within SBS. Section 1.5 indicates why common sense (CS) is not scientific, and explains how SBS relates to CS and how epistemic tasks are distributed between SBS and CS. Sections 1.3.4–1.6 indicate why Laudan's (1983) diagnosis of the demise of the SD problem was premature, why Laudan's proposal is a proposal for an epistemic-warrant and territorially broad demarcation, why the Mahnerian project seems to be a promising endeavour for seeking after an epistemic SD criterion (i.e., a criterion that determines an epistemic demarcation of science), how philosophers can defend an epistemic SD criterion and derived normative prescriptions, and how to demarcate, within that project, not only science (in the broad sense) from non-science and from pseudoscience, but also different types of sciences and of scientific fields (e.g., formal sciences, natural sciences, social sciences) from each other. Section 1.6 and Chap. 20 also indicate the width of practical applicability that an SD achieved through the completion of this project would enjoy.

1.2 Indirect Attempts at SD via the Demarcation of Cognitive Meaningfulness

Although attempts to define science date from at least Aristotle's *Posterior Analytics* (Laudan, 1983, p. 122),[6] the SD problem entered the academic field of the philosophy of science in the Twentieth century, and received a high degree of attention, as follows.

From the late 1920's, the logical positivists/empiricists of the Vienna and Berlin Circles were focused on establishing a criterion of cognitive (i.e., semantic) significance (cognitively significant/meaningful statements being those which are true or false). With the exceptions of the cases of analytically true and self-contradictory propositions, cognitive significance was made, in their account, coextensional with their notion of empirical meaningfulness. By doing this, they sought to dispel metaphysics from the realm of cognitively meaningful propositions. Their project was finding a criterion of meaningfulness which captured all non-metaphysical sentences and utterances that are considered to be cognitively meaningful, while

[4] Hansson used the term "science(s) in a broad sense" (2013, p. 64).

[5] In case there is a difference between being a property/criterion/aspect of science and being a property/criterion/aspect of scientificity, the inclusive disjunction "of science, or of scientificity" should be deemed implicit every time I write about properties, or criteria, or aspects, of science or of scientificity.

[6] Aristotle's SD is unsatisfactory. Firstly because of its ambiguity in his demands for scientific knowledge to be about what is universally or just "for the most part" (*hôs epi to polu*) constant (*Nicomachean ethics* (Aristotle, 2011), 1139a5–15; *Metaphysics* (Aristotle, 1961), VI.2, 1027a20; Parry, 2014). Secondly, because of the need to expand on his work, which for example did not encourage experimentation (Andersen & Hepburn, 2016).

discarding any metaphysical sentences and utterances. They tried to accomplish this with different successive criteria of empirical meaningfulness. And because they believed that the propositions of science were cognitively meaningful while the sentences and utterances (then) purported to be about metaphysics had no cognitive meaning,[7] they tried to use those different criteria of meaningfulness to discern science from metaphysics. This attempt to demarcate science from something that they considered to be non-scientific was, although indirect and partial, a scientific-demarcationist one.

Their theses on cognitive meaningfulness (CM)—verificationism, confirmationism, the criterion of translatability into an empiricist language thesis, etc.—suffered from several problems (some of which are summarised in Hempel, 1950, 1951, 1965) and were later abandoned. But even if any of these theses had been successful, that would have provided a criterion of CM, not a criterion of scientificity. CM is a necessary condition for propositions, and therefore for propositions within science, but any criterion of CM would be far from sufficient for our scientific-demarcationist purposes, because of at least three problems:

1. One problem with any criterion of CM as a tool for epistemic SD is that, by itself, a criterion of CM would indicate nothing about the epistemic status of at least many propositions: such a criterion could not be used to demarcate propositions whose acceptance is scientifically justified from any other type of propositions; while an epistemic SD criterion, by definition (see Note 2), should be capable of being used for that purpose.
2. A scientific demarcation must demarcate science from non-science, not broader or narrower categories. To illustrate: if we wanted to know which entities are Dobermanns and which entities are not Dobermanns, it would not help us much to look at which entities are animals and which entities are not animals, because being a Dobermann is much more specific than being an animal (a similar problem, although inverse, occurs with narrower categories, such as "social science").

   A problem with any criterion of CM as a tool for any type of SD is that the criterion would not uniquely capture the territory of science: for example, if all cognitively meaningful propositions were parts of science, then propositions of common-sense knowledge (also known as everyday, ordinary, or quotidian knowledge) would also be parts of science. This is a defeater because any SD that fails to capture *any* conceptual difference between common sense and science is not of our intended kind of demarcations (see Sect. 1.5).
3. A third problem with using any criterion of CM as a tool for SD is that science is more than just a set of meaning-bearing entities (such as theories, hypotheses, conjectures, or propositions): it also includes, among other components, methods, practices, institutions, and attitudes which give rise to those propositional elements.

[7] This would not include theoretical propositions within science, which are now sometimes called "metaphysical" (Chakravartty, 2017, pp. 19–20).

1.3 Direct Attempts at SD Since the Twentieth Century

1.3.1 Monocriterial Proposals

Some probabilists held that a theory is scientific if the mathematical probability of that theory according to the available total evidence is at least high. Lakatos categorised probabilism as a school within inductivist demarcationism (Lakatos, 1973, p. 33). Although Ritchie (1926) argued that the mathematical probability of any theory, given any amount of evidence, is 0, this did not deter the development of probabilistic confirmation theories, like the Bayesian one. In any case, this criterion alone cannot capture i) the non-propositional aspects in the notion of "science", nor ii) the notion of "scientific knowledge" as distinct from that of "common-sense knowledge", as it is not the case that every theory likely to be true is also a scientific theory: everyday, naïve, or folk (and therefore non-scientific) accounts (i.e., descriptions, analyses, or explanations), which can be theories at least in the broadest sense of the word, are often endowed with a great degree of probability given the available total evidence for a given individual or epistemic community. So it cannot be the case that high probability *per se* is a sufficient condition to make theories scientific.

The term "criterion of demarcation" ("*Abgrenzungskriterium*" in the original German) was coined by Popper in texts of *Die beiden Grundprobleme der Erkenntnistheorie* (*The Two Fundamental Problems of the Theory of Knowledge*) (2010) that circulated in 1930–33.

Popper's problem of SD was that of "finding a criterion which would enable us to distinguish between the empirical sciences on the one hand, and mathematics and logic as well as 'metaphysical' systems on the other" (1959/2002, p. 11). His solution was the in-principle falsifiability of sets of propositions (the theories) of empirical science, i.e., their capability "of conflicting with possible, or conceivable observations" (1962, p. 39), plus avoidance of abusing ad hoc definitions or ad hoc hypotheses to evade falsification (which is a conventionalist stratagem according to falsificationism) (ibid., p. 37). He conceived this as a necessary and sufficient composite criterion to indicate scientificity.

For our purposes, Popper's proposal suffers from the following problems:

(a) every falsified theory would be considered scientific (but to solve this one, it would suffice to require that a theory had been corroborated numerous times in different situations);
(b) the pragmatical problem that any theory falsifiable in principle but forever unfalsifiable in practice should be considered empirically scientific;
(c) in testing a theory, falsifiable assumptions that are not parts of that theory are also simultaneously tested. This is called contrastive holism. Because of this, what is falsified in the testing of a theory may be not a part of that theory;
(d) the problem, indicated by Bunge (1983, p. 201) and Hansson (2013, p. 74), that this criterion excludes those parts of scientific activity that are not concerned with the testing of theoretical constructs, such as some fact-finding or explorative

research or experimentation, and "[s]tudies [...] driven by open-ended questions" (ibid.), because it cannot be applied to them; and
(e) Popperian falsificationism also admits common sense as scientific (see Sect. 1.5).

To sum up, falsifiability, as a criterion for any demarcation of the kind intended here, can be excessively weak or excessively strong, and is neither a sufficient nor a necessary condition. Nonetheless, falsifiability might be fruitfully incorporated into a list of properties of scientificity used to demarcate science as indicated in Sect. 1.3.4.

Lakatos (1970) reformulated the problem of SD in terms of the theoretical progressiveness of scientific research programmes. These are sequences of successive theories that share a common hard core, "each theory having at least as much content as the unrefuted content of its predecessor" (ibid., p. 118). According to Lakatos, to be entirely progressive a programme must be both theoretically and empirically progressive, and if it is not entirely progressive, then it is degenerating. The theoretical progressivity of a scientific research programme comes entirely from the occurrence of a shift (within the programme) by which each new theory in the programme has an excess of empirical content over its predecessor theory, including making novel predictions. In this case, a theoretically progressive "problemshift"—or "theoryshift"—has occurred. Empirical progressivity comes entirely from the fact that at least some of these novel predictions are corroborated. In this author's proposal, good science is that of entirely progressive research programmes, while bad science is constituted by those degenerating programmes for which all "the novel predictions that they deliver turn out to be" falsified (Musgrave & Pigden, 2016). Regarding the demarcation of science (good and bad) from pseudoscience, according to Lakatos, "[w]e 'accept' problemshifts as 'scientific' only if they are at least theoretically progressive; if they are not, we 'reject' them as 'pseudoscientific'" (Lakatos, 1970, p. 118). So, Lakatos establishes two SDs (on the same continuum): one normative, between progressive (good) and degenerating (bad) science; and another one, between "scientific" and "pseudoscientific" problemshifts.

As indicated by Newton-Smith (1981), Lakatos's proposal is theoretically insufficient because it does not include some non-empirical aspects of science, namely conceptual aspects, which include questions of intelligibility, and of meaning. Newton-Smith correctly pointed out that "any model of science must leave room for the differential assessment of theories in terms of their power to avoid conceptual difficulties and not just in terms of their power to predict novel facts and explain known facts" (ibid., pp. 89, 208–36[8]). Lakatos's proposal is also pragmatically unsatisfactory due to the exclusively retrospective nature of its evaluations, which makes it impotent to offer "instantaneous" advice to scientists and philosophers about decisions to take and assessments to make in the present. So Lakatos's is not a satisfactory proposal to achieve our demarcatory goals, but the properties men-

[8] Newton-Smith referred here to theory choice, but the same can be applied to the demarcation between science and non-science.

tioned in it might be fruitfully incorporated into our list of properties that are typical of science (cf. Sect. 1.3.4).

Bunge (1983) rejected seven SD proposals: consensualism, the empirical content doctrine, the *success* view, formalism, falsificationism, methodism, and sociologism. Consensualism—the idea that "science is uncontroversial or at least aims at 'a consensus of rational opinion over the widest possible field'" (Ziman, 1979, p. 3)—is inadequate because (a) "every field of active basic research teems with controversy, and (b) consensus is at most an uncertain indicator of truth, not a goal of research. What is true is that science has means for settling controversies in the long run". The empirical content doctrine, which claims that there is no place in science for speculation, was refuted by the emergence of theoretical scientific disciplines. The *success* view—which "claims that for science only success counts"—misinterprets the goals of science, which are epistemic reliability, empirical adequacy, explanation, successful prediction, truth, depth, systemicity, and the like. The formalist view—which holds that a body of beliefs "is scientific only when it has been thoroughly mathematized"—is too wide in some cases (e.g., some formal economics) and too narrow in others (e.g., explorative science, experimental science, young science). We already (in this subsection) discussed Bunge's objections to falsificationism. Methodism—the idea "that the sole requisite for science is adopting the scientific method"—is problematic because "if every application of the scientific method were indeed a piece of scientific research, then testing the mental ability of atoms [...] would pass for scientific research provided certain precautions were observed". Bunge correctly discards sociologism—the proposal that science is what scientists do—, because "occasionally scientists indulge in nonscientific activities [...], or even straight pseudoscience" (1983, pp. 200–03). I agree with all of these Bungian assessments and reasons.

Authors working in the Strong Programme in the Sociology of Knowledge (e.g., Barnes, 1977, pp. 1–25; Bloor, 1991, p. 5) upheld a "relativist theory of knowledge" (Barnes & Bloor, 1982, p. 21). They proposed that, at least in the sociological sphere, scientific knowledge is, for any given sociohistorically-situated community, anything deemed scientific knowledge by that community. Barnes (1977, pp. 1–25) explicitly proposed a definition of "knowledge" for the sociological domain as "accepted belief", "publicly available, shared representations", "what people take to be knowledge", "knowledge claims", or what human beings produce when they "deploy their cultural resources to authentic tasks of explanation and investigation indicated by their interests". Bloor (1991, p. 5) proceeded in much the same way, indicating that "knowledge for the sociologist is whatever people take to be knowledge. It consists of those beliefs which people confidently hold to and live by", clarifying that "[o]f course knowledge must be distinguished from mere belief. This can be done by reserving the word 'knowledge' for what is collectively endorsed, leaving the individual and idiosyncratic to count as mere belief". This position, whose sociological merits we will not evaluate here, is philosophically unsatisfactory because it uses the word "knowledge" idiosyncratically: it explicitly defines "knowledge" for the sociological sphere as *any* belief widely accepted by and within any given society (i.e., a social belief), a class clearly different from the kinds of beliefs philosophers call (and should call) "knowledge", whether they consider knowledge to be factive (i.e., necessarily true) or not. (For more problems with the Strong Programme's positions, see Appendix B.)

Even a correct definition of "scientific theory" or of "scientific knowledge" would not be a complete definition of "science" as a whole, as it would leave aside practices, attitudes, social dynamics, and other components of science that are keys to its reliability.

In *Systematicity*, Hoyningen-Huene proposed a new monocriterial SD in which "[s]cientific knowledge differs from other kinds of knowledge, in particular from everyday knowledge, primarily by being more systematic [...] about the same subject matter[s]" (2013, pp. 14, 23) (his analysis also comprehends science as an activity, including methodological and sociological aspects of the way in which this knowledge is produced). By his own admission, Hoyningen-Huene defines systematicity in general and in the abstract in a tenuous and vague way, as a property of things that are "not purely random or accidental, [...] not chaotic, not arbitrary, not anyhow made or risen, not completely unmethodical, not completely unplanned, nor completely unordered. Rather, it embodies some kind of order" (ibid., 26). He correctly indicates that the possession of systematicity in general and in the abstract is a necessary but not sufficient condition for scientificity. He argues that a necessary and sufficient conjunction of conditions for knowledge or a theory to be scientific, or for a discipline to be a science, at time *t*, is the following: it possesses a sufficient amount of systematicity concretised in certain specific (scientific) ways indicated by him (2013), and, over a given period of time before *t*, that systematicity did not increase substantially less than the scientific systematicity of another contemporaneous discipline dealing with roughly the same subject matter (2013, 2018). Hoyningen-Huene affirms that science is demarcable because it is the sufficiently (and contextually the most) scientifically systematic knowledge/activity. This would be a very poor statement if it were not accompanied by a specification of the criteria demarcating scientific systematicity from other types of systematicity. And although Hoyningen-Huene takes important steps in that direction with his concretisations, his conception of scientific systematicity is used to refer to a set of more concrete properties that make up science but that we could and should identify more clearly (although they do confer to science a certain greater systematicity compared to other types of knowledge and of activities). He adopts an abstract concept of systematicity so vague that it allows him to classify and group as systematic all or most of the properties that have traditionally been considered to be key characteristics of science, which compels him to trivially yield the conclusion that the essence of science is scientific systematicity.

1.3.2 The Multicriterial Proposals of Hempel and Kuhn

Hempel advanced a proposal for multicriterial SD, arguing that philosophers should analyse differences in degree between theoretical systems in aspects such as a) "the clarity and precision" of their formulation; b) their "explanatory and predictive, power [...] in regard to observable phenomena"; c) "the[ir] formal simplicity"; and d) "the extent to which the theories have been confirmed by experiential evidence" (Hempel, 1951, p. 74; Ayer, 1959, p. 129). In order to achieve SD, we should take into account all of these aspects by incorporating corresponding properties into a list of properties that are typical of science (cf. Sect. 1.3.4).

Kuhn proposed an approximate SD based on five "standard" (1977, p. 322) scientific desiderata or epistemic values from a set of values which he deemed "canons that make science scientific" (ibid., p. 324). Each of these "values" can be descriptively thought of as being one or more properties, or "criteria for evaluating scientific belief" (1993, p. 338), and used in an SD criterion. Kuhn's selected group of "values" is:

1. theoretical accuracy of fit to nature—i.e., empirical adequacy—within an intended domain of application ("consequences deducible from a theory should be in demonstrable agreement with the results of existing experiments and observations" in its intended domain, and that agreement should be "not only quantitative [...] but qualitative as well"). He indicated that "predictive and explanatory powers" depend on this kind of accuracy;
2. consistency—internal and external, the latter being compatibility with other theories contemporaneously accepted—;
3. broad scope;
4. simplicity and the capability to bring "order to phenomena that [...] would [otherwise] be individually isolated and, as a set, confused"; and
5. fruitfulness of research findings of "new phenomena or previously unnoted relationships among those already known" (1977, pp. 321–23).

He mentioned a sixth epistemic value: the capability to be used to explain (ibid., p. 336).

All of these properties are worth incorporating into our list of scientificity properties (cf. Sect. 1.3.4). But Kuhn indicated (1977) that his list, by itself, is incomplete, and insufficient as a guide for good-theory choice because of at least six other problems, namely:

(i) "imprecision" of the "values", and the resulting lack of consensus among different legitimate assessments about the degree of presence of a given property in a given theory;
(ii) conflict between "values" in the absence of a consensus about how to "weigh" the respective degrees of importance of conflicting properties;
(iii) some of the "values" might select theories that we intend to discard;
(iv) we would need to specify what areas of a theory are more important for the satisfaction of a "value" if different areas of the theory satisfy that "value" to

different degrees (alternatively, we could aim to have a process to calculate the total degree to which a given theory satisfies a given "value");

(v) we would need to specify in what sense we mean the name of a "value", but this can be solved by using different words, phrases, or indexes for different senses; and

(vi) there can be legitimate disagreement over whether the possession of some of these properties can legitimately be deemed unnecessary for some fields to be considered scientific (ibid., p. 324).

These six obstacles to the use of these "values" in standardised decision procedures for theory choice specification are what compelled Kuhn to call these "values", i.e., "[c]riteria that influence decisions without specifying what those decisions must be" (ibid., p. 330).

Some of these six problems appear to be obstacles to the use of these "values" as criteria for SD, as they would be insufficient not only to determine theory choice, but also to specify scientificity and non-scientificity. Nonetheless, multicriteriality is not hopeless: further investigation, substantially extending Kuhn's list, and incorporating other provisions will help us overcome these obstacles, a possibility Kuhn admitted (ibid., p. 326).

We have reviewed the major attempts to demarcate science since 1930, and found that none of them succeeds. Popper's does not work because of the five problems mentioned when discussing his proposal. Lakatos's is unsatisfactory because it ignores conceptual aspects that are necessary to distinguish science from gibberish that predicts successfully. Hoyningen-Huene's account is incomplete because of its unspecificity. Hempel's proposal was just an outline of a project. And Kuhn's suggestion is only incipient and afflicted by the problems just mentioned.

1.3.3 From Multicriteriality to Anarchy and Capitulation

The lack of a solution to the SD problems is partly because the proposals offered could not account in full and exclusively for the extension of the pre-analytical notion of "science" applied by the scientific and philosophical international communities: it was always possible to offer examples of theories, disciplines, or practices that were not selected as scientific by the criterion or set of criteria proposed, although it was intended that they were (on the basis that they were already pre-analytically deemed scientific); or that were selected by the criterion or set of criteria suggested, but were actually intended to be excluded (on the basis that they were pre-analytically considered to be non-scientific).

Feyerabend proposed a plurimethodic[9] ("methodological anarchist") approach to science. In *Against Method*, he argued: "given any rule, however 'fundamental' or

[9] Following Bunge (2003, p. 180), I use "methodic" for things of, or relating, or pertaining, to methods (or sets or collections of methods), and I reserve the word "methodological" for things of, or relating, or pertaining, to studies of methods.

'rational', there are always circumstances when it is advisable not only to ignore the rule, but to adopt its opposite" (1993, p. 14). If that is the case, that does not imply that the violated rule is not, generally speaking, reliable, or at least sufficiently justified, and therefore advisable, so his is not a compelling argument for avoiding the inclusion of properties related to method in an SD, leaving the way open for the project presented in Sect. 1.3.4. The successful demarcation that will be born out of this project, in turn, will show that not just anything goes.

As a result of the failure to find a solution, Laudan declared that the problem of SD was in itself spurious, and rejected the idea that any epistemic criterion of SD could succeed (1983, p. 124). Despite its pervasive influence, all his diagnosis did was show that it is difficult to demarcate science from non-science by using a single simultaneously necessary and sufficient condition, as, taken individually, "none of the criteria which have been offered thus far promises to explicate the distinction" (ibid.), which is also shown here. We will get back to Laudan's dismissal in Sect. 1.6.

1.3.4 Rehabilitating the Multicriterial Approach

The failure of monocriterial SDs led other philosophers of science to propose multicriteriality, notably Bunge (1982, p. 372, 1983, 1984, 1991, pp. 245–47, 2003, pp. 259–62). Bunge's list of criteria of scientificity—which is partially used in Chap. 2—was further developed by Mahner (2007, 2013).

Thagard (1988) advanced a list of proposed properties of science, indicating that they belonged to the "conceptual profile of science". He contended that "we should aim for a list of features that are typical of science" (ibid., p. 159). In Kuhn's words, "[d]isciplinary characteristics permit its identification as […] science", locating every science "close to other scientific disciplines and at a distance from disciplines other than science. That position, in turn, is a necessary property of all referents of the […] term 'science'" (2000, p. 214). This would even allow science to be deemed a natural kind, while accepting a "moderate essentialism" (Mahner, 2007, pp. 521–22, 2013, pp. 39–40).

Consistently with the above, Mahner (2007, pp. 521–22, 2013, pp. 40, 42) proposed, to achieve SD, the project taken up in this work. It consists of: (a) the composition of a broad "list" (set) of accepted characteristics, conditions, or properties of science, or indicators of scientificity (most of them, by themselves, unnecessary and insufficient)[10] which might be collectively used to establish a demarcation

[10] Hansson (2017) mentions 15 proposed lists of SD criteria that can be considered. Fasce (2017) has called attention to the scarce amount of coincidence among these lists, worrying that it might be indicative of a lack of consensus among their authors. Nonetheless, there is also an almost complete lack of contradiction among them, which is due to the existence of a tacit generalised consensus in favour of the vast majority of the properties in these lists. The lack of coincidence is mostly due to variations in subject matter, or in approach, or in particular interest, or in degree of

between those theories, cognitive fields (CFs),[11] practices, etc. which are scientific and those which are not; and (b) calibrating the list of properties, and, in particular, ascertaining a boundary or "cut": the minimum score (measured in total aggregated amount of properties featured) that a theory, CF, practice, etc. would have to feature for it to be legitimately deemed scientific.[12] All this would be done by performing comparative analyses of a wide range of several clear cases of particular sciences and of non-sciences of different subtypes (natural sciences, social sciences, pseudosciences, non-pseudoscientific non-sciences, etc.).

That properties which, taken individually, are neither necessary nor sufficient conditions for scientificity can be used in a demarcation criterion should not puzzle the reader. It is not a problematic idea. It is a sensible procedure that we correctly employ even in everyday common-sensical reasoning. For example, in my home, eating while standing in the kitchen, making much noise while eating, having muddy feet while eating, and scratching vigorously while eating are, taken individually, neither sufficient nor necessary conditions to demarcate whether the eating agent in question is my dog or me. But when an agent, in a particular concrete event taking place in my home, displays three or all four of these properties, it is epistemically safe to conclude that the eating agent is my dog and not me. Sensible as this procedure is, it has never been attempted before using hundreds of properties of science for the demarcation of science, as is done here.

This proposal requires further elucidation and expansion with the following considerations. We need the "cut" value if we want to be able to say that the assessed unit is either scientific or non-scientific in the simplifying manner required by classical logic (because without the "cut", in many cases we would have to say, for example, that the assessed unit is 60 per cent scientific and 40 per cent non-scientific, and therefore both scientific and non-scientific at the same time and in the same sense, which would be more accurate but also more difficult to manage (see Appendix A)). The identification of the clear cases of particular sciences and of non-sciences will be made according to the assessments made by the current international philosophical and scientific consensus. Admittedly, philosophical consensuses are never unanimous, but the consensus about the extension of the term "science" is sufficiently extended and territorially-covering to serve as an acceptably firm base for intensional SD. In this way, we will be basing our intensional demarcation on a consensus about extension. A consensus should be deemed reliable to provide a base for philosophy only to the extent that that consensus can be shown to be the best available way to track (or the best available indicator of)

philosophical sophistication, or to differences in the degrees of importance adjudicated to different properties in incomplete lists, as none of these short lists is presented as exhaustive, presumably only those properties considered to be the most important being included in them.

[11] A "CF" is defined here, partially following Bunge (1984, p. 36), as "a sector of […] activity [purportedly] aiming at gaining [...] knowledge".

[12] We might be able to attain our intended SD without having to establish the "weight" (degree of importance) of each property relative to each other's, and without considering non-binary degrees of possession of each property.

epistemic reliability or at least success of outcomes, which is our ultimate grounding.[13] This course of action is consistent with the naturalistic contention that a criterion of scientificity "should be based on reference to criteria that are empirically observed to regulate the practices of science" (Ladyman et al., 2007, p. 33). The "cut" value for each type of unit could be taken to be the scientificity value of the lowest-valued clear case of a scientific unit of that type, provided that there were no clear case of a non-scientific unit of that type that had an equal or greater value. For example, the "cut" value for CFs would be established to be the scientificity value of the lowest-valued clear case of a scientific CF, provided that there were no clear case of a non-scientific field that had an equal or greater value. The properties can be classified under different categories, such as practices, aims, methods, logical aspects, conceptual aspects, products, and socio-institutional certification (cf. Erduran & Dagher, 2014; Irzik & Nola, 2014).

In this proposal, there are many different ways or manners in which a theory, CF, practice, attitude, etc. can be scientific or non-scientific, by possessing or failing to possess different subsets of (applicable) properties in our whole demarcatory set of properties of scientificity, and reaching or failing to reach the "cut" value (Mahner, 2007, p. 522, 2013, p. 40). Sciences of different subtypes are widely different to each other in several respects, and there is important variation among different stages of a science's development. That is why some properties of scientificity might be not equally relevant to, or might have not the same "weight" in, all sciences or across all stages of a science's development. Consequently, our demarcatory set of properties of scientificity could include properties that should be differentially computed (as indicated by the clear cases of sciences to be used as standards) for CFs of different subtypes or in different stages of their respective developments.

If this "multidimensionality" of science is the case, then one can legitimately ask why we would believe that the notion of "science" can be explicated as a single concept (let alone for now the problem of whether any such explicatum would correspond to a natural kind). The investigative project advocated in this work aims to find out whether there are sufficient reasons for holding such a belief.

If we do find out a set of (known) properties of scientificity that makes all scientific CFs, theories, practices, attitudes, etc. be scientific in the same sense (this is contingent, and therefore would have to be empirically discovered), then i) it would be possible and useful to compare very different fields, theories, practices, etc. with regard to their respective degrees of scientificity; and ii) a single total ordering of, say, CFs with regard to their respective degrees of scientificity would be possible as long as one knew the "weights" of all the properties in each CF's respective subset of possessed properties of scientificity (having considered the particular stage of development and the particular subtype of CF in question). i) and ii) would be

[13] Epistemic reliability is evidenced by success of outcomes, such as empirical adequacy, successful predictions, and functioning technological applications. All other properties of science (for example, properties related to method) are to work towards epistemic reliability, and should be deemed properties of scientificity, or not, with respect to the achievement of epistemic reliability.

possible even if these fields, theories, practices, attitudes, etc. were scientific (or non-scientific) in many different ways.

With the implementation of this method, i) some of the CFs consensually considered to be clearly scientific will not feature all the relevant properties listed in sufficient degrees, ii) some legitimate young science may be left outside of the domain identified as scientific, and iii) some pseudosciences will feature some properties of scientificity (cf. Kuhn, 2000, p. 214; Mahner, 2013, p. 39). We will probably be dealing with a vague concept: borderline cases will probably exist. These weaknesses are tolerable. That is why an epistemic justifiability profile of a CF, theory, practice, attitude, etc., if accomplished in this way, although it must be reasonable, might not reach the status of a clear-cut or definitive assessment. Given the acknowledged limits of the endeavour, it should not be considered necessary (at least in the short term) that the list of properties be exhaustive or unchangeable, provided that it conveys a sufficiently complete representation of science.

Also, because science is a historically and culturally situated activity, the degree of pseudoscientificity of a given discipline, theory, practice, etc. can vary socially and historically (Thagard, 1978, p. 227, 1988, pp. 169–70; Pigliucci, 2013, p. 18; Hansson, 2013, p. 74; Mahner, 2013, pp. 34–36; Hoyningen-Huene, 2013, p. 206), because the epistemic circumstances—from which we take our reasons for believing—alter as we (appear to) learn more, or forget what we have (appeared to) learn, or as we justifiably believe differently.

1.4 Descriptivity, Normativity, and Territoriality in SDs

The idea that reliability is key to demarcate science from pseudoscience is defended by Hansson (2013). Two conceptual points against Hansson should be made. Firstly, he equates reliability with being "as close to correctness as can currently be achieved" (ibid., p. 66), but neither correctness nor being "as close to correctness as can currently be achieved" is equivalent to reliability. Some descriptions made using the Ptolemaic model of the Solar System are reliable to a certain extent, but now we know that model to be incorrect and we have a more correct model (although correctness can be used as evidence of reliability). Secondly, Hansson equates reliability with epistemic warrant (ibid., pp. 67, 70), but, again, these two notions are not equivalent. Epistemic warrant (justification) is even weaker than reliability, as it does not necessarily lead to reliable results (for example, in Gettier cases).

Unlike Hansson (ibid., pp. 73–5), this work maintains the hypothesis (subjected to eventual falsification or improvement) that science (or at least our best examples of it) is currently, in general, mature enough for properties related to method to be included into a general and timeless definition of the term "science".

The demarcation of science from non-science, and the demarcation of science from pseudoscience, are SDs based on properties that can be correlated to higher or lower degrees of epistemic reliability or success of outcomes. For example, biology, as a science, features certain properties that art lacks and that make biology be epistemically more reliable than art; and astronomy features certain properties that astrology lacks and that make astronomy predict better than astrology.

The SDs of this type are descriptive and not normative, because all they require are solely descriptive propositions indicating that a property is or is not sufficiently possessed or that a condition is or is not sufficiently met, and that there is a correlation between that property or condition being sufficiently possessed or met and higher or lower degrees of epistemic reliability or success of outcomes. I call these SDs epistemic-warrant SDs (EWSDs).[14] Therefore, a definition or demarcation of science, or of non-science, or of pseudoscience, need neither rely on nor imply any commitment to values or norms.[15] Any such definitions or demarcations would only become normative upon the inclusion of, or their conjunction with, normative content.

EWSDs can be used to derive normative prescriptions, for example of the form "We should believe *x* and disbelieve *y*".

Several positions on EWSD can be combined with different positions on what Boudry (2013, pp. 79–81) calls territorial demarcation, which "is concerned with a classification of knowledge, or a division of labor between different disciplines, and not with epistemic warrant" (justification)—unless some fields offer less justification than others. For example, if physics and chemistry offer the same amount of epistemic warrant, then demarcating between them is not an exercise in EWSD, but a mere exercise in territorial demarcation (and because physics and chemistry are both sciences, it is also an intra-SD).

Particular instances of EWSD (e.g., concluding that ethology is a scientific CF) can be territorial in the sense that the particular sets of epistemic properties on which they are based can be exclusively or better epistemically suited to respective definite fields. For example, ethology is epistemically reliable in part because it has to take into account learned behaviours, but this property is not well suited to other scientific CFs, such as chemistry.

Some of the most broadly comprehensive and usual CFs mentioned in the academic bibliography on philosophy of science are: common sense (CS), natural sciences, social sciences, psychology, "hard" sciences, "soft" sciences, technological sciences, empirical sciences, the humanities (including philosophy, human history,

---

[14] This name was chosen because scientific epistemic warrant (scientific justification) is the best way available to us to arrive at propositions that can be the content of our most reliable beliefs on issues about reality that are difficult for us to accurately evaluate (i.e., issues which are the purview of science). Scientific epistemic warrant is achieved when CFs, theories, practices, attitudes, etc. are sufficiently endowed with the above-mentioned properties that can be correlated to higher degrees of epistemic reliability (cf. Mahner, 2013, p. 37).

[15] Even if we use the prefix "pseudo-", that does not necessarily imply that we are making a normative assessment, because the fact that something is false may be good, bad, or neither good nor bad. An example of it being good is if Socrates had drunk false poison. Incidentally, this is also an example of a situation in which it would even have been good to present something false as the real thing. All concepts that are denoted by words are normative in the trivial sense that there are rules about how we ought to use those words, but Resnik (2000, p. 251) is wrong in conflating this sense with other senses of "normative".

etc.), formal sciences, mathematical sciences, complexity sciences, and SBS, which includes all the other ones mentioned except for CS (see Sect. 1.5). Each of these CFs encompasses certain disciplines, attitudes, practices, and intended kinds of outputs (except for CS, which is not disciplined enough to encompass disciplines), so different fields could be characterised by respectively different sets of properties (cf. Kuhn, 1977, p. 331). Some fields completely engulf other, smaller fields (i.e., subfields); for example, entomology is a subfield of zoology, which is in turn a subfield of biology.

Certain disciplines (e.g., behavioural neuroscience, which includes biology and psychology) cross over putative territorial borders, making the identification of those borders problematic. Also, there is interdependency, continuity, overlapping, and cross-fertilisation between these CFs, which is for example evident between philosophy (including metaphysics) and science (Boudry, 2013). Quine proposed that our total science, "from the most casual matters of geography and history to the profoundest laws of atomic physics or even of pure mathematics and logic" (1961, pp. 42–43), belongs to one and the same "fabric" or "web", in Laudan's words, that all of it is "cut from the same epistemic cloth" (1983, p. 124, 1990). Because of this, authors such as Bunge (1983), Hoyningen-Huene (2013), Hansson (2013, pp. 62–64), and Boudry (2017) proposed to demarcate science as SBS (i.e., what Hansson called the "community of knowledge disciplines"), a superdomain that—like the German-language notion of "*Wissenschaft*"—includes all the "systematic and critical investigations [appropriately] aimed at acquiring the best possible understanding" of reality, including nature, ourselves, our societies, our physical and thought constructions (Hansson, 2017), and anything that interacts with those spheres and leaves detectable traces of those interactions (Boudry, 2011, 2017; Boudry & Leuridan, 2011). Some issues about reality are difficult for us to accurately evaluate, which is why we need systematicity and criticality beyond CS in order to (better) sort them out. SBS includes every CF able to regularly lead an epistemic agent or community to hold beliefs that are sufficiently, and contextually the most, reliable, or at least the most intersubjectively[16] epistemically justified (and better epistemically justified), on issues that are difficult for us to accurately evaluate (cf. Hansson, 2013, p. 70). (The possession of this ability is a high-level property/criterion of scientificity that does not specify how to arrive at the class of beliefs required. We humans have discovered that sufficient possession of certain low-level properties leads to the possession of the high-level property. The possession of all these low-level properties implies the possession of good conceptual, methodic, empirical, and socio-institutional credentials, which are specified by respective low-level properties.) SBS is the superdomain of all the disciplines that provide the most reliable and fertile—or at least the better founded and most credentialed—non-common-sense knowledge. So SBS encompasses all disciplines falling under all the territorially narrower legitimate notions of science, it includes all propositional

[16] Nagel (1961, p. 449) defined "intersubjective validity" as the property of having publicly accessible supporting evidence.

parts of the humanities, and all CFs that satisfy the necessary and sufficient condition of having the aforementioned ability in their respective domains (cf. Boudry, 2013, p. 93). Hansson (2013, 2017) contends that the CF of SBS is the most adequate one to consider in order to understand the epistemic contrasts implicated in the demarcation of science from pseudoscience. Also, generally speaking, the CFs within SBS seem to be increasingly interdependent.

1.5 Science and Common Sense

The superdomain of SBS excludes intellectual activities (and their fruits) that have been seen at times as contiguous with science, e.g., common-sense reasoning, common-sense knowledge-seeking, common-sense knowledge acquisition, and common-sense knowledge (also known as everyday, ordinary, or quotidian knowledge). Any SD that fails to capture *any* conceptual difference between CS and science is not of our intended kind of demarcations, because our pre-analytical as well as our philosophically and scientifically informed notions of science imply that there is at the very least *some* difference between science and CS, and therefore the extensions of those notions of science do not include common-sense knowledge and practices. Common-sense knowledge and practices can sometimes be as empirically determined, justified, and reliable, or more so, than scientific ones, and they can be so in ways similar to those of science. CS crosses over to science, and is used in science. Science can discover facts that are also discoverable by CS, and the converse is true to a certain extent. Some elements of CS can change in time, or be simultaneously different in different epistemic communities, some of these changes and variations being merely due to CS becoming loaded with scientific knowledge or with beliefs of other kinds. CS can be protoscientific in the sense that it is previous to science and, in conjunction with certain other provisions, gives rise to science. None of these facts count against science, or preclude us from conceptually separating those components. "[T]he scientist is more careful" and "systematic" than "the common [hu]man" (Quine, 1957, pp. 5–6). This increased degree of care is the reason why science is more epistemically potent than CS (Haack, 2007, pp. 24–25, 94–109). "[T]he scientist is more narrowly focused on issues of truth and objectivity and, in the hope of contributing to these goals, clearer" (Hylton & Kemp, 2019).

Some authors go further, indicating that science departs from CS and natural language to the point of being "unnatural" (e.g., Newton-Smith, 1981, pp. 210–12; Cromer, 1993; Wolpert, 1993; McCauley, 2000). This position is compatible with my position expressed in the immediately previous paragraph.

Regarding cases of CFs, knowledge, practices, etc. whose correct categorisation appears to be difficult to ascertain (e.g., the professional investigation of crimes or accidents), we may find that they fall wholly either on one side or the other of the line (territorially) demarcating science from CS, but very close to it; or that some of them fall within a third, "middle zone", having the respective "weights" of characteristics that are typical of each of these two realms in sufficiently similar aggregated amounts. None of these alternatives would be a fatal problem for the type of SD proposed here.

In any case, an epistemic community or agent needs science only to the extent that it cannot discover or be sufficiently certain (in a reliable or at least intersubjectively-justified epistemic manner) just by using CS (but it cannot always be sufficiently certain by using science). Common-sense knowledge on issues that are easy for us to evaluate can be as certain as scientific knowledge on issues that are difficult for us to evaluate, or more so. Boudry (2017) correctly indicates that the line of demarcation between CS and science is relative to a) the specific endowments of a particular epistemic community—its cognitive and perceptual capabilities—; and b) "the standards of evidence and explanatory depth we demand in any given situation"; to which should be added c) the degree of precision sought.

1.6 Chapter Conclusions

We should not give complete credit to Laudan's, 1983 obituary, as the developments reviewed here show that it was premature. But Laudan's paper inadvertently gives weight to arguments in favour of the territorially broad SD. Indeed, Laudan's conclusion is that, because (a narrower territorial) demarcation appears to be unreachable, what ought to be demarcated instead is reliable knowledge from non-reliable knowledge; well-founded beliefs from those that are not; claims with conceptual and empirical credentials in their favour, from claims lacking such credentials; and beliefs that are heuristically fertile from beliefs that are not. That is a proposal for an epistemic-warrant and territorially broad demarcation. And it turns out that Laudan's conclusions are precisely some of the premises leading other authors to propose to demarcate science broadly! (although Laudan's proposal was territorially too wide: it referred to "reliable knowledge", but this would include common-sense knowledge, which is non-scientific, as argued in Sect. 1.5).

It seems that we can epistemically demarcate SBS from other intellectual activities and outputs. Both types of SDs, territorial and epistemic-warrant, are seemingly feasible of SBS, and from non-SBS. By using the territorially broad SD, in which all of SBS *is* "cut from the same epistemic cloth", one could defend an epistemic SD criterion, and derived normative prescriptions.

Regardless of the advantages that a territorially broad SD could yield, any demarcation of the CF of SBS could be narrowed down by the introduction of different sets of constraints in order to adapt it to demarcate respectively different scientific subfields. More specifically, different subdivisions of the CF of SBS could be demarcated naturally, i.e., in a non-arbitrary way that corresponds to actual philosophical and scientific reflection and practice. For example, adding the requirement that the justified or reliable beliefs be only about nature would demarcate the subfield of the natural sciences. And it appears that respective EWSDs are feasible for such territorial subdivisions. This is why using the concept of SBS will allow us not only to demarcate SBS from non-SBS and from pseudoscience, but also to demarcate different types of sciences and of scientific CFs within SBS (e.g., formal sciences, natural sciences, social sciences, human history) from each other, so as to not overlook the important specificity, with respect to properties, of such subfields.

This work agrees with Resnik (2000) that "any particular definition of science may be evaluated in terms of the consequences of accepting that definition" (ibid., p. 262). This is due to the fact that the use of a more reliable definition of the term "science" is more likely to better help us serve our interests and reach our goals, whatever they be. This kind of evaluation is appropriate only within the limits of, and with respect to, a theoretical definition of the term "science" that must feature other content, as success is only an indicator of scientificity, and neither a sufficient nor a necessary condition for it, as history clearly shows.

However, Resnik (2000) also contends that our definition of the term "science" should vary by context, as different definitions of science may have different effects on the degree to which we accomplish different goals. But analytical explicative definitions are, by their very nature, inherently theoretical, not practical. This does not impair our capacity to "explore and evaluate different points of view and [...] discuss new or controversial ideas" (cf. ibid., p. 263). Resnik's position stems from the fact that he conflates a definition of the term "science", on the one hand, with different degrees of stringency that we might need to apply in order to get the best results in different situations, on the other. His confusion is generated by the fact that he writes about practical decisions regarding what should be admitted in different contexts, and he conflates admissibility (or utility) with scientificity. Moreover, his contention that "[s]ince the decision to call a human activity 'science' has important social, political, and practical implications, one should not expect that an apolitical theory or definition of science should be able to solve the demarcation problem" (ibid., p. 265) is simply a non sequitur, as it is very common for apolitical theories to have important social, political, and practical implications, e.g., our apolitical theory of disease etiology leads us to vaccinate children.

Pigliucci (2022) complains that my

> project [...] runs into significant difficulties for a number of reasons. First, [...] Fernandez-Beanato wishes for more precision than is likely possible, [...] aiming at a quantitative "cut value" on a multicriterial scale that would make it possible to distinguish science from non-science or pseudoscience in a way that is compatible with classical logic. It is hard to imagine how such quantitative estimates of "scientificity" may be obtained and operationalized. Second, the approach assumes a unity of science that is at odds with the [...] emerging consensus in philosophy of science that "science" (and, similarly, "pseudoscience") actually picks a family of related activities, not a single epistemic practice. Third, Fernandez-Beanato rejects Hansson's (and other authors') notion that any demarcation criterion is, by necessity, temporally limited because what constitutes science or pseudoscience changes with our understanding of phenomena. But it seems hard to justify Fernandez-Beanato's assumption that "Science ... is currently, in general, mature enough for properties related to method to be included into a general and timeless definition of [the term] ["]science["]" [...].

I still say that (above, in Sect. 1.4). Pigliucci's critique deserves adequate answers. Let us provide them in corresponding order: 1) The contention that a high level of demarcatory precision is likely impossible has not been sufficiently substantiated.

Furthermore, this book is an example of operationalisation of my proposed demarcatory technique that yields quite precise boundaries. 2) Given that the abstract, high-order definition of the term "science" proposed in this chapter, together with the demarcatory technique proposed in this work are so far the only successful demarcatory proposal that we know of, this work is proof that science has threads going through its entire extension, and therefore that science is more monolithic than a family-resemblance concept. 3) On the contrary: what is hard to imagine is how most of the methodic properties listed in this book could be non-scientific at any time in any place. Could it ever, in any place, be scientific to ignore the majority of the good evidence at hand, or to use epistemically spurious strategies to avoid being critical about your ideas? I think that the answer is quite clearly "No".

The demise of the SD problem was prematurely diagnosed because of the insufficiency of the tried proposals, and due to scepticism about multicriteriality's prospects. Philosophers of science should not have capitulated so soon. It is our hope that this chapter has indicated a clear way ahead in the form of an executable project to achieve SD.

References

Andersen, H., & Hepburn, B. (2016). Scientific method. In E. N. Zalta (Ed.), *The Stanford Encyclopedia of Philosophy* (Summer 2016 ed.). Stanford University. https://plato.stanford.edu/archives/sum2016/entries/scientific-method/

Aristotle. (1961 [4th century BCE]). *Metaphysics*, Books I–IX, Books X–XIV, Tredennick, H. (trans.), Loeb Classical Library. Cambridge, MA: Harvard University Press.

Aristotle. (2011 [4th century BCE]). *Nicomachean ethics* (2nd ed.), Irwin, T. (trans.). Indianapolis: Hackett Publishing.

Ayer, A. J. (1959). *Logical positivism*. The Free Press.

Barnes, B. (1977). *Interest and the growth of knowledge*. Routledge.

Barnes, B., & Bloor, D. (1982). Relativism, rationalism and the sociology of knowledge. In M. Hollis & S. Lukes (Eds.), *Rationality and relativism* (pp. 21–47). MIT Press.

Bloor, D. (1991). *Knowledge and social imagery* (2nd ed.). University of Chicago Press.

Boudry, M. (2011). *Here be dragons: Exploring the hinterland of science*. Ghent, Belgium: Ghent University, Faculty of Arts and Philosophy. https://biblio.ugent.be/publication/1191286

Boudry, M., & Leuridan, B. (2011). Where the design argument goes wrong: Auxiliary assumptions and unification. *Philosophy of Science, 78*(4), 558–578.

Boudry, M. (2013). Loki's wager and Laudan's error. On genuine and territorial demarcation. In M. Pigliucci & M. Boudry (Eds.), *Philosophy of pseudoscience: Reconsidering the demarcation problem* (pp. 79–98). The University of Chicago Press.

Boudry, M. (2017). Plus ultra. Why science does not have limits. In M. Pigliucci & M. Boudry (Eds.), *Science unlimited? The challenges of scientism* (pp. 31–52). University of Chicago Press.

Bunge, M. A. (1982). Demarcating science from pseudoscience. *Fundamenta Scientiae, 3*, 369–388.

Bunge, M. A. (1983). *Treatise on basic philosophy, Vol. 6: Epistemology & Methodology II: Understanding the world*. Reidel.

Bunge, M. A. (1984). What is pseudoscience? *Skeptical Inquirer, 9*, 36–46.

Bunge, M. A. (1991). A skeptic's beliefs and disbeliefs. *New Ideas in Psychology, 9*(2), 131–149.

Bunge, M. A. (2003). *Philosophical dictionary* (Enlarged ed.). Prometheus Books.

Chakravartty, A. (2017). *Scientific ontology: Integrating naturalized metaphysics and voluntarist epistemology*. Oxford University Press.

Cromer, A. (1993). Uncommon sense: The heretical nature of science. *Science, 265*, 688.

Erduran, S., & Dagher, Z. R. (2014). *Reconceptualizing the nature of science for science education: Scientific knowledge, practices and other family categories*. Springer.

Fasce, A. (2017). What do we mean when we speak of pseudoscience? The development of a demarcation criterion based on the analysis of twenty-one previous attempts. *Disputatio. Philosophical Research Bulletin, 6*(7), 459–488.

Fernandez-Beanato, D. (2020a). Cicero's demarcation of science: A report of shared criteria. *Studies in History and Philosophy of Science Part A, 83*, 97–102.

Fernandez-Beanato, D. (2020b). The multicriterial approach to the problem of demarcation. *Journal for General Philosophy of Science, 51*(3), 375–390.

Fernandez-Beanato, D. (2020c). Correction to: The multicriterial approach to the problem of demarcation. *Journal for General Philosophy of Science, 51*(3), 391.

Fernandez-Beanato, D. (2021). Feng Shui and the demarcation project. *Science & Education, 30*(6), 1333–1351.

Fernandez-Beanato, D. (2023). Apocalyptic pseudoarchaeology. *Skeptical Inquirer, 47*(2), 48–51.

Feyerabend, P. (1993). *Against method* (3rd ed.). Verso.

Haack, S. (2007). *Defending science—within reason: Between scientism and cynicism*. Prometheus Books.

Hansson, S. O. (2013). Defining pseudoscience and science. In M. Pigliucci & M. Boudry (Eds.), *Philosophy of pseudoscience: Reconsidering the demarcation problem* (pp. 61–78). The University of Chicago Press.

Hansson, S. O. (2017). Science and pseudo-science. In E. N. Zalta (Ed.), *The Stanford Encyclopedia of Philosophy* (Summer 2017 ed.). Stanford University. https://plato.stanford.edu/archives/sum2017/entries/pseudo-science/

Hempel, C. (1950). Problems and changes in the empiricist criterion of meaning. *Revue Internationale de Philosophie, 41*(11), 41–63.

Hempel, C. G. (1951). The concept of cognitive significance: A reconsideration. *Proceedings of the American Academy of Arts and Sciences, 80*, 61–77.

Hempel, C. G. (1965). *Aspects of scientific explanation and other essays in the philosophy of science*. Free Press.

Hoyningen-Huene, P. (2013). *Systematicity: The nature of science*. Oxford University Press.

Hylton, P., & Kemp, G. (2019). Willard van Orman Quine. In E. N. Zalta (Ed.), *The Stanford Encyclopedia of Philosophy* (Spring 2019 ed.). Stanford University. https://plato.stanford.edu/archives/spr2019/entries/quine/

Irzik, G., & Nola, R. (2014). New directions for nature of science research. In M. R. Matthews (Ed.), *International handbook of research in history, philosophy and science teaching* (pp. 999–1021). Springer.

Kuhn, T. S. (1977). Objectivity, value judgment, and theory choice. In *The essential tension: Selected studies in scientific tradition and change* (pp. 320–339). The University of Chicago Press.

Kuhn, T. S. (1993). Afterwords. In P. Horwich (Ed.), *World changes. Thomas Kuhn and the nature of science* (pp. 311–341). MIT Press.

Kuhn, T. S. (2000). *The road since structure: Philosophical essays 1970–1993*. University of Chicago Press.

Ladyman, J., Ross, D., Spurrett, D., & Collier, J. G. (2007). *Everything must go: Metaphysics naturalised.* Oxford University Press.

Lakatos, I. (1970). Falsification and the methodology of scientific research programmes. In I. Lakatos & A. Musgrave (Eds.), *Criticism and the growth of knowledge* (pp. 91–195). Cambridge University Press.

Lakatos, I. (1973). The theological nature of scientific standards. In M. Matteo (Ed.), *For and against method* (pp. 31–40). The University of Chicago Press.

Laudan, L. (1983). The demise of the demarcation problem. In R. S. Cohen & L. Laudan (Eds.), *Physics, philosophy and psychoanalysis* (pp. 111–127). Springer.
Mahner, M. (2007). Demarcating science from non-science. In T. Kuipers (Ed.), *Handbook of the philosophy of science: General philosophy of science – Focal issues* (pp. 515–575). Elsevier.
Mahner, M. (2013). Science and pseudoscience. How to demarcate after the (alleged) demise of the demarcation problem. In M. Pigliucci & M. Boudry (Eds.), *Philosophy of pseudoscience: Reconsidering the demarcation problem* (pp. 29–44). The University of Chicago Press.
McCauley, R. N. (2000). The naturalness of religion and the unnaturalness of science. In F. C. Keil & R. A. Wilson (Eds.), *Explanation and cognition* (pp. 61–85). The MIT Press.
Musgrave, A., & Pigden, C. (2016). Imre Lakatos. In E. N. Zalta (Ed.), *The Stanford Encyclopedia of Philosophy* (Winter 2016 ed.). Stanford University. https://plato.stanford.edu/archives/win2016/entries/lakatos/
Nagel, E. (1961). *The structure of science: Problems in the logic of scientific explanation*. Harcourt, Brace & World.
Newton-Smith, W. (1981). *The rationality of science*. Routledge.
Parry, R. (2014). Episteme and Techne. In E. N. Zalta (Ed.), *The Stanford Encyclopedia of Philosophy* (Fall 2014 ed.). Stanford University. https://plato.stanford.edu/archives/fall2014/entries/episteme-techne/
Pigliucci, M. (2013). The demarcation problem. A (belated) response to Laudan. In M. Pigliucci & M. Boudry (Eds.), *Philosophy of pseudoscience: Reconsidering the demarcation problem* (pp. 9–28). The University of Chicago Press.
Pigliucci, M. (2022). Pseudoscience and the demarcation problem. In *Internet Encyclopedia of Philosophy*. https://iep.utm.edu/pseudoscience-demarcation/
Popper, K. (1959/2002). *The logic of scientific discovery* (Trans. *Logik der Forschung*. Repr.). Hutchinson.
Popper, K. (1962). *Conjectures and refutations: The growth of scientific knowledge*. Basic Books.
Popper, K. (2010 [1930–33]). Die beiden Grundprobleme der Erkenntnistheorie. Mohr Siebeck.
Quine, W. V. O. (1957). The scope and language of science. *British Journal for the Philosophy of Science, 8*, 1–17.
Quine, W. V. O. (1961). *From a logical point of view* (2nd ed.). Harper & Row.
Resnik, D. B. (2000). A pragmatic approach to the demarcation problem. *Studies in History and Philosophy of Science Part A, 31*(2), 249–267.
Ritchie, A. D. (1926). Induction and probability. *Mind, 35*, 301–308.
Thagard, P. (1978). Why astrology is a pseudoscience. *PSA, 1*, 223–234.
Thagard, P. (1988). *Computational philosophy of science*. MIT Press.
Wolpert, L. (1993). *The unnatural nature of science: Why science does not make (common) sense*. Harvard University Press.
Ziman, J. (1979). *Reliable knowledge*. Cambridge University Press.

Chapter 2
A Survey of the Properties of Science for Multicriterial Demarcation

Abstract This chapter explains what classification of the properties of science is the most appropriate for a multicriterial intensional epistemic demarcation of science. It then presents the most comprehensive list of properties of science that has been compiled in the history of the philosophy of science.

Keywords Science · Scientific demarcation · Non-science · Pseudoscience · Scientificity

2.1 Introduction

As a method to reach an accurate SD, the multicriterial approach is promising because many mutually compatible accounts of what science is focus on different real aspects of science. While many philosophers have noticed this in the past, none so far had systematically developed this approach enough.

In order to develop it, this work starts by compiling those properties of science that have been deemed key in the literature on SD (in this chapter). Because we have realised what works in science, many of the properties of science are also norms or rules of it, and are, generally speaking, sufficiently well-applied and adhered to in the sciences. Many of these properties are abstractions or generalisations, and can take different specific forms in different domains or contexts. For example, there are different types of unification that can be separately considered: in some scientific fields (e.g., physics), unification can be realised mathematically, while in others (e.g., biology), it might take the form of accounting for diverse lines of evidence; and in yet other fields unification might take the form of theorising a single underlying ontology.[1] These different forms

[1] The list of properties in this chapter is not exhaustive in this respect, because a full discussion of a single property and of all the different forms that it might take would take pages, and is therefore practically impossible, within the extension of this work, to do that for each property listed. A work

Parts of this chapter were previously published in Fernandez-Beanato (2021).

D. Fernandez-Beanato, *A Working Scientific Demarcation*, Synthese Library 519, https://doi.org/10.1007/978-3-031-99005-2_2

are usually mutually compatible, and therefore they might be present in one and the same field. The possession of a property in a given particular form might be domain-specific, and some of the properties might be more or less important in different domains, contexts, or stages of a science's development. This diversity partially explains why authors have emphasised different dimensions and different properties of science when talking about different scientific fields, and why many of them have failed to notice the possibility of multicriterial demarcation.

Then, in order to see whether this multicriterial approach is viable, I check a) whether it rationally reconstructs and tracks the widespread academic consensus about the extension of the term "science", b) whether it yields an ordering demarcating the set of the clear cases of science from the set of the clear cases of non-science, and, c) in case it does, whether this order includes a noticeable differentiating gap between those two sets.

It is impossible for one author, in a first-order manner, to survey the entirety of each field analysed here and determine whether each one of them meets each of the compiled criteria of science. This work, instead, is based on meta-analyses, in that it relies on hundreds of other specialist authors who have written on particular fields and identified those fields' particular properties (for demarcationist and other purposes). This is a big part of my data. When a given property is attributed to a field by authors who are experts in that field, or on the philosophy of that field, writing in widely used or cited canonical texts on those subjects, said attribution is counted as a reliable datum. It is impossible to survey everything said in the literature about the scientificity or non-scientificity of any field, but I aim to convey representative samples.

Chapters 3–19 each consider a particular field, including a wide range of clear cases of particular sciences and of non-sciences of different subtypes (natural sciences, social sciences, pseudosciences, non-pseudoscientific non-sciences, etc.), determining whether the field has the properties. If any adjudication of a property turns out to be exceptional in the sense that it is legitimately controversial, then it can be eliminated without endangering the overall conclusions, as there will still be enough adjudications to substantiate both the conclusions of each case study and the overall conclusions. I record the evidence of attributions of properties, with citations and quotations of passages where authors make said attributions. The analyses provide scores roughly corresponding to the degree of scientificity of each field. The scoring method is explained in Sect. 2.3.

Many different sets of criteria can be demarcatory of many different concepts denoted by the term "science". But not all of these sets correspond minutely to, and therefore demarcate well, the real phenomenon of science. Given the epistemically-centred notion of "science" presented in Sect. 1.4, it follows that the most accurate intensional SD is an epistemic[2] one. This work looks for an *accurate* or *correct*

---

of the extension of this entire work could be written on all the properties as they are possessed by just one of the several sciences analysed here.

[2] "Epistemic SD" is defined in Note 2 of Chap. 1.

intensional SD, i.e., an intensional epistemic SD that captures the actual extension of the term "science", and nothing else, in a non-arbitrary manner.

The qualification of "epistemic" for the kind of demarcation wanted here is necessary because non-epistemic intensional demarcations of concepts denoted by the term "science" can exist. For example, a single criterion based on the property of being done in London, taken as a necessary and sufficient condition, would yield the following intensional non-epistemic SD: "'Science' is whatever is done in London". If a property is epistemically irrelevant, then it is irrelevant to, and consequently useless for, any epistemic demarcation. Therefore, epistemically irrelevant properties are not epistemically demarcatory of science at all. Non-epistemic attempts at intensional SD are incorrect because they ignore science's intensionally definitory property of having an epistemic guarantee or evidential basis that is, generally speaking, epistemically safer than that of non-science.

If this is applied to our example, it becomes clear *why* the definition "'Science' is whatever is done in London" is obviously incorrect as an intensional SD: because, science being what it actually is, the property of being done in London is intensionally irrelevant to something being scientific, and it is non-contingently irrelevant to deciding whether something is to be deemed scientific (although it might be contingently relevant to that decision, simply because a disproportionate percentage of all human science has been done in London).

An intensional epistemic SD that corresponds to our existing ideas about the actual extension of the term "science" has eluded philosophers so far because science is a complex phenomenon composed of many epistemically-relevant "dimensions", facets, or aspects: it includes methodic, propositional, pragmatical, psychological, attitudinal, social, axiological, institutional, and other types of epistemically-relevant components.

In the literature on SD, authors identify many properties that they use as criteria of science, or of scientificity. As some authors have noted, the properties of science can be classified as of different kinds or types. There is no standard classification of (the relevant) properties of science in the philosophy of science. Laudan wrote of a basic twofold distinction: good "empirical and conceptual credentials" (1983, p. 125). Pigliucci (2013, pp. 22–3) wrote about "theoretical understanding" and "empirical knowledge". Erduran and Dagher (2014), building on a taxonomy developed by Irzik and Nola (2014), proposed the following categories: practices, aims, methods, products, and socio-institutional certification. The former two suggestions seem to be too concise to be useful as a full-fledged taxonomy, while the latter is arguably overgrown (as, for example, the differentiation between practices and methods is not readily clear).

Sometimes, some of the different epistemically-relevant aspects of science are naturally fused. For example, a single property having to do with the practices of institutionalised scientific communities could include inseparable sociological, institutional, pragmatical, and axiological aspects. Consequently, the aspects are not always neatly separated into different properties.

For its purposes, in its analysis this work distinguishes between different kinds or types of properties/criteria. Firstly, it divides the properties of science into those that are irrelevant to an epistemic SD, and those that are relevant to an epistemic SD. As mentioned above, an example of the former is the property of being done in London, which is intensionally irrelevant to determining whether something is scientific, and is not epistemically demarcatory of science at all. Therefore, I ignore them.

Not all of the properties presented here are equally useful for demarcation, and not all of them are computed in the same way in our assessments of scientificity. This is because there are properties that are essential, or typical, but are neither exclusive nor specific enough of some of the things of which they are essential or typical properties, and therefore the possession of those properties is not very helpful for demarcation. For example, having DNA is essential, or necessary, of adult humans, and being over 1 m tall is typical of them, but the possession of either of these properties will not help us demarcate adult humans from non-humans.

This is asymmetrical, in that the lack of the property does help us demarcate (in our examples, entities lacking any DNA must be non-human, and entities of less than 1 m of stature, epistemically speaking, are probably not adult humans). This is the case of those properties that are typical of science and that are also possessed by other fields. For example, the property of having propositional content is common to almost all fields analysed in this work (except perhaps for art).

This work calls these the "non-specific" properties of science. They are generic properties of epistemically-good CFs (but not exclusive to them). Some of these shared properties are important to science. Many of the properties of science that we think are important to it being science are also properties possessed by other CFs, such as common sense, and are present in everyday practices. This becomes evident when one considers that there are principles of inquiry that are generally epistemically good to apply, both in common sense and in science. For example, both in everyday life and in doing science, people should not disregard evidence. Looking down the street to see if a car is coming is not doing science, and the knowledge obtained by looking is not scientific knowledge. So it is patent that not disregarding evidence is not demarcatory of science in particular. In the same manner, being concerned with truth and falsity is important for science, but also for buying a car, and presenting a cognitively meaningful discourse is important both for science and for everyday human relationships.

Nonetheless, asymmetrically, if a unit of analysis features disregard for evidence or for truth and falsity, or fails to present a cognitively meaningful discourse, this disregard or failure would be demarcatory in that it would count against identifying said unit as scientific.

Properties whose possession is demarcatory can be less important than some of those that are not. For example, being typically clothed is obviously not as important to being human as having DNA, but the property of being typically clothed can help us demarcate humans from elephants, while the property of having DNA cannot. In the same way, facticity (our property number 4, see below) is more important to science than being able to describe causal mechanisms in a highly specific theoretical manner, but the latter can help us demarcate science from factual common sense, whereas the mere possession of facticity cannot. (Some of the properties of science might be more or less important in different domains, contexts, or stages of a science's development.)

In light of all this, our list of epistemically-relevant properties that are typical of science (in Sect. 2.2) is divided into two portions. Its first portion lists properties that are not exclusive to science (i.e., they are "non-specific" to science). An example of a non-specific property is the property of having propositional content. This property is shared by the sciences, common sense, the pseudosciences, religions, etc. Therefore, possessing these properties is demarcatory of science only "negatively", i.e., the possession of any one of those properties does not count in favour of a unit of analysis being scientific (it should not, and in my account it does not, increase a given unit of analysis' scientificity score), but the lack of them (if they are epistemically-relevant to the given unit of analysis) counts against the unit of analysis being scientific. The second portion of our list is composed of properties that are highly specific to science. These are both negatively and positively demarcatory, in the sense that possessing any one of them counts towards a given unit of analysis being scientific (i.e., it adds to its scientificity score), and lacking any one of them (if they are epistemically relevant to the unit of analysis being examined) subtracts from that score (i.e., it counts against the unit of analysis being scientific). An example of a specific property of science is the property of having adopted the general scientific method as the main strategy for dealing with cognitive problems.[3]

This classification of properties can be grasped via the following example. The property of being in London is mostly irrelevant to determining whether Josephine is an adult giraffe. Among the properties that are relevant to determining it, the property of being an animal is relevant but non-specific, so Josephine not being an animal would count against it being an adult giraffe, but knowing that Josephine is an animal would not be very helpful to establish whether it is an adult giraffe. On the other hand, the relevant property of being a live animal over 4.2 meters tall is specific enough such that lacking it should count against it being an adult giraffe, and having it should count in favour.

Furthermore, an original classification of properties of science is provided here that divides the set of all the (epistemically-relevant) properties listed into five topical main categories, having to do, respectively, with: i) practices, methods, and techniques; ii) formal properties of propositions, hypotheses, theories, and arguments; iii) contents of problems, propositions, hypotheses, and theories; iv) epistemic attitudes and values; and v) epistemic properties of propositions, hypotheses, and theories. For example, featuring blind testing is a property of the first type; the possession of internal logical consistency (non-self-contradiction) is of the second type; in-principle falsifiability is of the third type; featuring a general outlook including the

[3] Because the properties that are epistemically relevant must be legitimately applicable, some properties that are epistemically relevant to some sciences (and that because of that are epistemically relevant to (broad) science in general, and are listed here as epistemically relevant) might be epistemically irrelevant to other sciences. For example, properties regarding empirical testing might be inapplicable, and therefore irrelevant, to the assessment of the formal sciences such as logic and mathematics.

ethos of the free search for truth is of the fourth type; and general self-correctiveness is of the fifth type. I recognise that the 220 properties could conceivably be classified in alternative manners. As this work shows, the classification I devised was sufficient for achieving the intended intensional SD. My categorisation is not exclusive in that certain properties could be listed under more than one category; in those cases, the properties in question are listed only under the first appropriate category. Each of the five types of properties is in turn bifurcated into a list of those that are specific to science and a list of those that are not. This yields a total of 10 sub-lists that together make the complete overall list.

Different units of analysis (e.g., CFs, theories, research programmes, practices, attitudes, testimonies) can be assessed with regard to their lack or possession of properties of science. All units of analysis that are accepted as scientific by the consensus of the global philosophical and scientific communities are identified as scientific by them because those units of analysis possess good scientific credentials (qualifications). A unit of analysis has good scientific credentials tout court if and only if it possesses enough of the specific properties of science (which is probably impossible if it lacks too many of the non-specific properties of science).

Different types of good scientific credentials are possessed by a unit of analysis when it possesses enough specific properties of science of correspondingly different types. For example, a unit of analysis possesses good propositional scientific credentials if and only if it possesses enough of the propositional and specific properties of science (which is probably impossible if it lacks too many of the propositional and non-specific properties of science) ("propositional" meaning that these properties of science are most directly related to science's propositional content, rather than being most directly related to other aspects such as methods, attitudes, etc.).

To illustrate the different epistemically-relevant aspects of science, consider the contrasts between science and Lysenkoism, a pseudoscientific view of genetics, biological evolution, and agriculture that was put forward in the Soviet Union in the 1920s. Lysenkoism is a good example of a pseudoscience, of how the different epistemically-relevant aspects of science can be lacking, and of their importance, for at least four reasons, namely: a) because Lysenkoism clearly fell short in several of these aspects; b) because that failure had many of the great ethical, political, and social implications that can be relevant to the main SD problem, and shows how great the importance of the problem is; c) because Lysenkoism was explicitly claimed to be scientific by its advocates; and d) because Lysenkoism has been extensively commented on by philosophers of science, and explicitly taken as a paradigmatic (in the everyday, not the Kuhnian sense of this word) example of pseudoscience by them.

In the specific case of Lysenkoism, if we analyse it qua CF, we can distinguish, among the properties of science that it mostly lacked (corresponding to epistemically-relevant aspects of science in which it fell short), properties of method, properties of propositions, and psychological, attitudinal, social, and institutional properties.

Let us examine Lysenkoism with reference to these different epistemically-relevant aspects/types of properties, comparing Lysenkoism side by side with science.

Regarding properties of method: generally speaking, science typically satisfies a great many of its methodic desiderata. Among them are the transparency/publicity of its methods (Bunge, 1960, p. 14; cf. Lack & Rousseau, 2016, p. 40; Tuomela, 1985, p. 212), an abundance of rigorosity (cf. Lilienfeld, 2016, p. 268; Gruenberger, 1964, p. 1414; Lawson, 2007, p. 4; Bunge, 2013, pp. 1–48), carefulness (Lilienfeld, 2016, pp. 268–70; Simon, 1969, p. 414), being painstaking (Simon, 1969, p. 414), using control groups in testing (Lack & Rousseau, 2016, pp. 46–7), and recognising the value of statistics (Thagard, 1988, pp. 162–3, 2011, p. 27). Instead, Lysenkoism was secretive about its methods, it advocated moving fast from hypothesis to belief and implementation (Borinskaya et al., 2019), and it denied the usefulness of control groups and statistics (Faulk, 2013).

Regarding properties of propositions: science is empirically successful, i.e., it works generally well in describing, predicting, and allowing us to manipulate, phenomena. This is a general feature of science, which permits a higher degree of success than that permitted by mere common sense and traditional techniques. During the 1950s and 1960s, the so-called Third Agricultural Revolution (also known as the Green Revolution), based on science, is credited with having saved over one billion people from starvation. In contrast, Lysenkoism was an epistemo-technological disastrous failure that made crop yields decline where it was applied, contributing to the Great Chinese Famine, which killed millions of people and might have been the biggest famine in the history of humanity by number of deaths.

Regarding psychological, attitudinal, social, and institutional properties: psychologically, science requires an in-principle willingness to accept that one might be wrong (Bunge, 1960, pp. 6–10; Kitcher, 1982, pp. 32–3; Lack & Rousseau, 2016, pp. 20, 22–3, 31, 47; Lilienfeld, 2016, p. 271; Gruenberger, 1964, p. 1415). Attitudinally, sociologically, and institutionally, science benefits from free and open debate, discussion, exchange of ideas (cf. Bunge, 2013, p. 26; Tuomela, 1985, pp. 210–1, 213–4), and peer-review (Park, 2003; Lilienfeld, 2016, pp. 268–9; Coker, 2001, p. 6; Ladyman et al., 2007, p. 28) (all of which, serving as checks and safeguards, could show one that one is wrong, or how one could improve one's propositions). Science also needs intellectual freedom and autonomy (Merton, 1973 [1942], pp. 257–9; Tuomela, 1985, pp. 216–8; Mahner, 2013, p. 38). Lysenkoists, on the other hand, charged that the propositions of their scientific critics were "bourgeois", "imperialist", and "fascist" views, pushed by "people-haters", "saboteurs", "perverts", obscurantists and representatives of backwardness. Lysenkoism was supported by the central governments of two major countries, and won politically (in the USSR in 1948 and in the People's Republic of China in 1949), becoming for a time the "law of the land" in those countries and in Poland and Czechoslovakia. Lysenkoism's scientific opponents were banished from academic institutions and in some cases imprisoned or killed.

In all these manners, Lysenkoism was epistemically and scientifically wrong not only in its methods and in its propositional content, but also in other wider, and even scientifically "external", pragmatic aspects. This facilitated and reinforced the continued adherence to, and defence of, its misguided propositions and methods.

It is not surprising that an intellectually unchallenged group got it wrong. This example has been an extreme one for the sake of clarity. Lysenkoism is easy to classify (correctly) as non-scientific because it was wrong in all of these general aspects of science. But a non-scientific unit of analysis could fare more or less well in some of them. It is not necessary for a unit of analysis to be so full-fledgedly non-scientific in so many aspects in order for it to be correctly deemed non-scientific.

This multiplicity of epistemically-relevant aspects of science has the consequence that, even if a succinct definition of SBS can be given (as was done in Sect. 1.4), such a definition is difficult to apply consensually in practice without the specification of its many more concrete, low-level properties of different types. That is why Chap. 1 indicated that the first step to take in order to achieve SD, which is to be executed here, is the composition of a list of properties of science or indicators of scientificity. This chapter introduces that list of properties, via a comprehensive survey of the properties of science that have been discussed in the academic philosophy of science literature.

There has been a pause in the debate about SD. Paradoxically, this situation coexists with an immense corpus of mostly mutually consistent philosophical sources about demarcation. My method was to take a very broad list of high-quality works on demarcation, and to catalogue the different properties that have been proposed as criteria of scientificity in that corpus.

The following is a list of all the high-quality authors who have published proposed lists of SD criteria that I have been able to find. In Chap. 1 were reviewed those of Hempel (1951, p. 74), Popper (1959/2002, p. 11, 1962, p. 39), Lakatos (1970, p. 118), Kuhn (1977, pp. 321–22, 336, 1996, p. 185), and Hoyningen-Huene (2013). Hansson (2017) highlights: Langmuir ([1953] 1989), Merton (1973 [1942]), Gruenberger (1964), Dutch (1982), Bunge (1982), Radner and Radner (1982), Kitcher (1982, pp. 30–54), Hansson (1983), Grove (1985), Thagard (1988), Casti (1990), Glymour and Stalker (1990), Derksen (1993, 2001), Vollmer (1993), Ruse (1996, pp. 300–6), and Mahner (2007). Fasce (2017) adds Tuomela (1985), Schick and Vaughn (1995), Beyerstein (1996), Coker (2001), Park (2003), Lilienfeld et al. (2012), and Lack and Rousseau (2016).[4] Other important works advocating particular properties of science are those of Russell (1949), Bunge (1960, 1984, 2003, 2013), Simon (1969), Oreskes (2004), Shelton (2004, Chap. 9), Ladyman et al. (2007), Boudry and Buekens (2011), Carey (2011), Law (2011, n.d.), Thagard (2011, pp. 26–9), Boudry and Braeckman (2011, 2012), Boudry (2013a, 2013b), Fales (2013), Goode (2013), Hansson (2013), Ladyman (2013), Mahner (2013), Nickles (2013), Pigliucci (2013), Shermer (2013), Ruse (2013), Lilienfeld et al. (2015), Lilienfeld (2016), Fernandez-Beanato (2017), Harker (2017), Blancke et al. (2018, 2019), and Mukerji and Ernst (2022). I use all of the above.

---

[4] Of the lists considered by Hansson, I have ignored most of the properties mentioned in Langmuir ([1953] 1989), because it refers to "pathological/sick science", which, although caused by self-delusion, is arguably still within the domain of science. Of the lists considered by Fasce, I have ignored those that are self-published (Skelton, 2011) or that were written for non-academic primary purposes (Overton, 1982).

Many of the properties of science surveyed here were presented in negative form in these sources, as reasons why some units of analysis are pseudoscientific, or were presented as properties of good scientific theories, focusing on the propositional aspect of science.

It is a commonly-held view that there is hitherto no commonly accepted unitary account of what science is (e.g., Brigandt, 2013, p. 225). This view is true of explicit, articulated, explicated, and approximately complete accounts. As noted in Chap. 1, some authors (e.g., Fasce (2017)) have called attention to the small amount of overlap among the above-mentioned lists of properties of science—most of which are considerably short (consisting of 10 criteria or less)—, worrying that it might be indicative of a lack of consensus among their authors. Indeed, these authors do not always explicitly advocate all the same criteria. This occurs in the face of a generalised consensus about the extension of the term "science" (see below in this section) because the properties of science number in the hundreds. They are hundreds because science is a very complex phenomenon.

But the idea that there is hitherto no commonly accepted unitary account of what science is, is misleading if it is taken to mean that there is no commonly accepted fuzzy or vague intension (notion) of "science" that can serve as our explicandum. An almost complete lack of dispute among these authors with regard to the properties of science,[5] and the fact that many of the different authors tacitly incorporate many of the other authors' criteria into their own analyses, point to the fact that there *is* a large measure of compatibility (non-contradiction) about the intension of "science".

There is, indeed, an existing and philosophically-informed agreement about the intension of "science", in that the vast majority of well-informed philosophers of science, other specialists in science studies, and scientists who are clearly (uncontroversially) scientists would agree that the vast majority of the properties that are listed here are properties of science. This would yield us a consensus about a multifaceted intension that determines, in turn, the existing consensus about the extension of "science". This is the best explanation for the phenomenon of ready recognition, i.e., the fact that the vast majority of well-informed philosophers of

[5]This is the case with some exceptions which have been left out of our list—notably, the demands for: i) exclusively natural ontology and problematics, ii) a priori methodological naturalism, and iii) scientific realism—, the first two of which are analysed in Chap. 5 (regarding a priori methodological naturalism, the approach taken here puts no a priori epistemic territorial constraints on what content can be scientific: it admits that, in principle, science could be about anything, i.e., the content of its propositions does not have to be exclusively about something in particular, for example, the natural world; it does not even have to be about reality, as it is possible to discuss counterfactuals scientifically). In fact, the 63 authors used to make the demarcatory list are only a sample of the numerous authors who have written on this topic, the vast majority of whom do not contradict their peers otherwise. The rest of the apparent differences among the lists are due to variations in subject matter, or in approach, or in particular interest, or in degree of philosophical sophistication, or to differences in the degrees of importance adjudicated to different properties in incomplete lists, as none of these lists is presented as exhaustive, presumably only those properties considered to be the most important being included in them.

science and scientists who are clearly (uncontroversially) scientists are able to extensionally differentiate between almost all scientific and non-scientific practices, disciplines, theories, attitudes, modes of procedure, etc., and do so or would do so in much the same way (cf. Pigliucci & Boudry, 2013, p. 2; Pigliucci, 2013, p. 11; Boudry, 2013b, p. 83).

Boudry (2022) worries that there are more than one potential explanations for said ready recognition:

> Several philosophers have noted [...] that [...] both scientists and philosophers are in large agreement (and sometimes unanimity) [...]. Virtually everyone agrees that homeopathy, creationism, phrenology, graphology, Freudian psychoanalysis and astrology are not good sciences, while general relativity, evolutionary biology and the germ theory of disease most certainly are. Even those philosophers who would refrain from using the term "pseudoscience" will generally arrive at the same judgements: the first class of theories and practices have hardly any epistemic merit, and should not be funded or taught in the classroom, while the second class have and should. This near-unanimity could lead one to believe that perhaps deep down everyone is tacitly relying on the same theoretical virtues (Schindler, 2018) and we just have not fully explicated them yet. But this is not necessarily true. It may also be the case that different criteria happen to converge on the same classifications, or that everyone just follows the judgement of the same relevant experts. Yet another possibility is that the defects from which the aforementioned theories suffer are completely different in every case, though they are all serious enough to justify the label "pseudoscience". (Boudry, 2022, p. 88)

This deserves a response. First, it is worth noting that it is possible to give a single intension of "science" and a coherent list of properties of science, on which the vast majority of authors agree, to explain the extensional agreement on which units of analysis are scientific and which are not. This book does all that. Second, of course my account is, as Boudry says, not *necessarily* true in the logical sense (i.e., it is not tautological). Third, the idea that "everyone just follows the judgement of the same relevant experts" is easily dismissed because we find that there is compatibility and agreement among a wide-ranging and heterogeneous group of philosophers of science and other specialists in science studies who have had different influences, read different literature, studied different areas of science, and who certainly do not all follow the same epistemic authorities. Fourth, Boudry's hypothetical convergence (i.e., virtually everyone agreeing on the same extensions while having very different intensions/sets of criteria in mind) seems highly unplausible, which is why the burden of proof would be on anyone trying to substantiate that hypothesis. Fifth, different low-level criteria (in fact, hundreds of them, as listed in Sect. 2.2) do converge on the same classifications/extensions, as showed in this work. Sixth, non-sciences, including pseudosciences, are indeed correctly classifiable as such because of different epistemic defects and other types of properties.

Agreement about extension across a wide variety of cases is prima facie evidence for agreement about intension. The overlapping extension is explained by an overlapping intension: most specialists agree on what is science and what is not science because most of them mostly agree on what scientificity is.

All this, in turn, leads to the conclusion that the main problem of SD has already, in a sense, been solved, although an explicative integrated account of that solution has not yet been given, and doing so is this work's goal. Therefore, this work searches for a specific type of intensional epistemic SD: one that corresponds to, rationally reconstructs, explains, and perhaps has the potential to explicate, our existing philosophical and scientific consensus regarding the extension of the term "science".

When I say that there exists a consensus about SD, I do not mean that there are no grey areas or contested cases; I only mean that the international philosophical and scientific consensus deems most cases clear. In between clear cases of scientificity and of non-scientificity, there might be a third, "middle zone", inhabited by units of analysis which have the respective "weights" of characteristics that are typical of each of these two realms in sufficiently similar aggregated amounts. Therefore, although this work's hypotheses are confirmed and corroborated via tests against 17 examples of fields, after applying the demarcatory method proposed here to many more units of analysis, we might end up with a grey or borderline area. "Science" is probably a vague concept, in that borderline cases probably exist (for example, String Theory is an interesting contested case to evaluate). Even if this happens, the demarcation achieved here would still be instructive, in that it a) would provide us with an explication of the intension that is behind our already-existing extensional identification of clear cases of scientificity and of non-scientificity (together with a justification of that identification), and it b) would properly identify borderline cases as such and provide an explication/justification of that identification.

This work presents a list of 220 epistemically-relevant properties of science, of different types, that are key criteria for SD. They are the low-level properties that concretise the abstract definition of SBS given in Sect. 1.4. A unit of analysis can be shown to be scientific by showing that it possesses enough of the specific properties of science, and that it does not lack too many of the non-specific properties of science (in other words, by showing that it possesses good scientific credentials); and if the unit possesses enough specific properties of science of different types and does not lack too many non-specific properties of science of those types, that shows that said unit is scientific in the corresponding different aspects of science (in other words, that it possesses good scientific credentials of those different types).

This work's **main hypothesis** is that the sole application of the scoring system proposed in Sect. 2.3 (based on the possession or lack of the relevant properties on the list) determines, in each case, an SD of CFs, theories, practices, attitudes, etc. that corresponds to the extensional demarcation of the current philosophical and scientific consensus regarding the extension of "science" in the broad sense.

The truth value of this main hypothesis does not depend on a) the acceptance of the higher-level definition of the term "science" proposed in Sect. 1.4 or on b) the truth value of our above-mentioned hypothesis about the existence of an overlapping intension of "science", because this work's proposal could in-principle be, and is, successful in intensionally demarcating science in accordance with the current consensus on the extension of "science", independently of what the authors contributing the properties had in mind.

That of fields is only one of many possible units of analysis that are analysable along these lines, other possible units being theories, research programmes, practices, attitudes, testimonies, etc. This work only tests its hypotheses against fields. This choice had to be made because the amount of work required to test this work's hypotheses against different units of analysis is immense. Fields were selected as our target unit of analysis because they are the kind of units offering more advantages. As Mahner puts it:

> Choosing fields of knowledge as a starting point allows us to consider the many facets of science, namely that it is at the same time a body of knowledge and a social system of people including their collective activities. It also takes into account that science is something that did not come into existence ready-made, but rather has developed over several centuries from a mixed bag of ordinary knowledge, metaphysics, and nonscientific inquiry. In other words, it allows us to consider not just the philosophy of science, but also the history, sociology, and psychology of science. Moreover, the study of epistemic fields requires that we also analyze their components. In this way, the smaller or lower-level units, such as statements, theories, methods, and so on, can be included as necessary components in our account. (Mahner, 2013, p. 36)

Regarding philosophical consensuses on clear cases of particular sciences, many CFs are nearly universally accepted as scientific in the broad sense, e.g., the formal sciences, the natural sciences of physics, chemistry, astronomy, and biology, and, with less certainty, psychology and sociology. The fields of analytic philosophy and human history (when properly conducted) are also nearly universally accepted as sciences in the broad sense.

As part of the strategy for demarcation proposed here, this work subjects the favoured traits (properties) detailed in this chapter to be empirically found in consensually clear (agreed-upon) cases of sciences. And it does find them there. This is not surprising, since the presence of such properties in successful and reliable scientific fields is part of the evidence taken into account to advocate for those properties in the first place. This involves a certain virtuous circularity, in that the analyses of clear cases not only served for the purposes of composition and calibration of the list of SD criteria, but also for the empirical "confirmation" or "disconfirmation" of those criteria as such. This way, I derive a normative theory of the quiddity of SBS in part from analyses of existing successful sciences.

Also, several clear cases of *non*-scientific fields feature prominently in the academic scientific and philosophical literature. Genethliac astrology has been used profitably as a clear (and paradigmatic, in the everyday, not the Kuhnian sense of this word) case of a non-scientific and pseudoscientific CF according to a consensus shared by philosophers, scientists, and historians of science (Thagard, 1978, p. 223; Grim, 1990a, p. 15; Pigliucci, 2013, p. 17). Parapsychology (at least the type that resorts to paranormal explanations) is another CF commonly considered to be non-scientific and pseudoscientific (Grim, 1990b, pp. 183, 186; Sommer, 2014, p. 42); and creation "science" is often considered to be a prime example of a non-scientific and pseudoscientific CF (cf. Mc Mullin, 1993, p. ix; Boudry, 2011, p. 4) by both philosophers and scientists. The same is true of homeopathy (e.g., Tuomela, 1987, pp. 93–4, 99; Smith, 2012; Cleland & Brindell, 2013, pp. 183–4; Hansson, 2013, pp. 69–70; Koertge, 2013, p. 178; Ladyman, 2013, pp. 49, 53; Mahner, 2013, pp. 30, 35; Pigliucci & Boudry, 2013, p. 3; Baran et al., 2014, p. 41). The contention that common sense is not scientific is correctly justified in Chap. 1. And, there is no substantial doubt that fields such as art and religion are non-scientific.

Several case studies, including analyses on all of these CFs qua CFs, are presented here, confirming the main hypothesis. To take stock: there is a consensus about the extension of the term "science", and this work makes explicit the complex intension that corresponds to it, finally solving the main problem of SD.

This work does not claim to present or to list the intensional properties of science in their simplest forms; rather, the expressions of properties are taken as they are articulated in the literature. Because of this, some of the items mentioned here as a single property are in fact i) clusters that could conceivably be further decomposed into conjunctions or disjunctions of simpler properties; or ii) are composed of a main property or properties with some preferable specifications indicated. This work does not claim that there is no redundancy in its list of properties, as overlapping is bound to occur in the expressions of different properties. No effort was made to eliminate as much redundancy as possible because this work aims to give a compendium of scientificity criteria as they are presented in the literature. It does strive for completeness. After making a limited number of judgement calls to exclude implausible proposals, I included the vast majority of the properties advocated by the authors whose works I used here, certainly listing all properties that have been the object of repeated advocacy in the philosophy of science academic literature. In the future, the list will probably have to be slightly modified or enlarged as a consequence of further research.

Full justification of the inclusion of each property in the list is not given, but citations of the philosophers of science who have appealed to these properties for SD are provided. The referencing of these properties to the works of different authors who defended them should not be construed as an endorsement of any of these authors' other positions or stances. The inclusion of a property in the list does imply that I agree that that property is typical of, or advisable in, science.

2.2 The Demarcatory List

The following is the most comprehensive list of properties of science that has been compiled in the history of the philosophy of science.

2.2.1 Properties that Are Non-specific to SBS

Arguments justifying the classification of each of the following 84 properties as non-specific to science are not given, because these statuses are rather uncontroversial. To exemplify with a randomly chosen property: it is fairly obvious that objectivity (our property 2) is neither exclusive nor specific enough of science, being also a property of epistemically-good everyday (common-sensical) thinking and of good judicial thinking.

2.2.1.1 Non-specific Properties Related to Practices, Methods, and Techniques

1. Theoretical findings are done by combining data with educated guesses (hypotheses) (cf. Bunge, 2013, p. 28).
2. Objectivity, i.e., adaptation of ideas to facts via the participation of, or commerce with, said facts in the process (Bunge, 1960, p. 10). Science is objective ("at least in the sense of intersubjectivity") (Tuomela, 1985, p. 212).
3. Is concerned with truth and falsity (cf. Ladyman, 2013, pp. 45–59).
4. Facticity: Science starts from facts, adheres to them to a certain degree, transcends some of them, and comes back to them (Bunge, 1960, pp. 11–2).
5. Does not display indifference to facts (Coker, 2001, p. 1).
6. Does not make up bogus "facts" (ibid.).
7. Features "confirmations and disconfirmations, in contrast to pseudoscience, which is typically oblivious to such matters" (Thagard, 1988, p. 162). The possession of this property requires the possession of property 54.
8. Features rational debate (cf. Bunge, 2013, p. 26).
9. Avoids abusing ad hoc definitions to evade theoretical falsification (which is a conventionalist stratagem) (Popper, 1959/2002, pp. 19–20). Does not move the semantic goalposts (Law, 2011, Chap. 4).
10. Avoids abusing ad hoc hypotheses to evade theoretical falsification (which is a conventionalist stratagem) (Popper, 1959/2002, pp. 19–20, 59–62, 131, 438; 1962, p. 37). Science does not "overuse of ad hoc hypotheses designed to immunize claims from falsification" (cf. Lilienfeld et al., 2012, pp. 21–3; Lilienfeld et al., 2015, p. 7). It avoids ad hoc rescues (Carey, 2011, pp. 120–1). It does not give only ad hoc explanations (Tuomela, 1985, p. 225).
11. Does not use *built-in subterfuge*. The testing of a theory is not so arranged that the theory can only be confirmed, never disconfirmed, by the outcome (Hansson, 1983, cited in Hansson, 2017).
12. Does not use unfounded immunisations (Derksen, 1993, pp. 23–5; Boudry, 2011, 2013a, 2019, 2022; Boudry & Buekens, 2011; Boudry & Braeckman, 2011, 2012; Blancke et al., 2018, 2019; Mukerji & Ernst, 2022).
13. Considers its disconfirmations, misses, and incorrect predictions (cf. Lilienfeld, 2016, p. 269). It does not ignore all findings that contradict its conclusions (Coker, 2001, p. 4). It does not ignore conflicting evidence (ibid., p. 1). The doctrine's theorists and practitioners do not dismiss unresolved problems as "exceptional cases" (Kitcher, 1982, pp. 48–9). Keeps track of failures (cf. Bunge, 2013, p. 18). "Failures are searched for and studied closely" (Coker, 2001, p. 6). The doctrine's theorists do not refuse to follow up on unresolved problems (Kitcher, 1982, p. 48). The possession of this property requires the possession of property 54.

14. Reactive dynamicity: Does not tend to remain stagnant in the face of contradictory evidence (Lilienfeld et al., 2015, p. 11): its hypotheses and theories are or would be changed due to confrontation with recalcitrant adverse empirical or other evidence (cf. Tuomela, 1985, p. 228).
15. The phenomena it studies do not appear "only under certain vaguely specified but vital conditions (such as when no doubters or skeptics are present; when no experts are present; when nobody is watching; when the 'vibes' are right") (Coker, 2001, pp. 4–5).
16. Does not refuse to let others see its data for independent evaluation (ibid., p. 7).
17. Leaves room for speculations to be superseded (Radner & Radner, 1982, p. 92).
18. Does not have special methods through which any wanted result in a wide range can be yielded (Derksen, 1993, pp. 30–32).
19. Checks results (Coker, 2001, p. 2).
20. Checks effectiveness (cf. Bunge, 2013, p. 18).
21. Does not stack or rig arguments, for example: "Heads I win, tails you lose"; and does not use loaded questions (Dutch, 1982, p. 11).
22. Limits its epistemic relativism (ibid., p. 12).
23. Looks for correlations (Thagard, 1988, pp. 162–6), and uses correlation reasoning/thinking (Thagard, 2011, p. 27): inferring "that two things or events are causally related from the fact that they are correlated with each other". "Correlation thinking should be construed broadly to encompass generalization from examples, statistical reasoning, and other kinds of induction. Inference to the best explanation qualifies, since it postulates causes on the basis of their ability to explain and predict observed regularities" (Thagard, 1988, pp. 162–3). Correlation reasoning/thinking "applies statistical methods to find patterns" (Thagard, 2011, p. 27).
24. Avoids illicit causal inferences, such as from a "simple correlation between A and B" or from a "concomitant variation between A and B" (Carey, 2011, pp. 113–5). In the case of a simple correlation,

    > there seem to be at least two plausible explanations for the data[...] that do not involve any sort of causal link [...]. The first is that the correlation is just a *coincidence*. If we look at a number of groups by profession, we may now and then find one where there is a significantly greater number of people born under a particular [astrological] sign, particularly if we restrict our investigation to groups that are none too large. [... (This is)] predictable "clumping."
    >
    > [...] It may be that the researchers who undertook the study have presented us with only one small part of their overall data, the part that appears to confirm the possibility of a causal link. Or it may be that [...] they have inadvertently pruned away just enough data—say, by excluding certain subjects—to lend support to the idea of a correlation.
    >
    > The explanation for a correlation need not be coincidence or even fudging, inadvertent or otherwise. Frequently, correlations are explained by some third factor which suggests a possible indirect link between the correlated factors. (ibid., p. 114)

    Regarding concomitant variations, the problem

    > is that an enormous number of entirely unrelated things tend to vary in very regular sorts of ways. Over the past ten years there has been a dramatic increase in the popularity of

> country-western music. At the same time there has been a corresponding increase in the cost of a loaf of bread. What is the explanation here? A genuinely baffling causal link? Some overlooked third factor? The most likely explanation is that we have managed to pick two completely unrelated trends that happen to be going in the same direction at the same time. (ibid., p. 115)

25. Lack of appeal to mysteries: Does not appeal to unfounded non-mysteries (cf. Tuomela, 1985, p. 228; Carey, 2011, pp. 110–112), i.e., presenting a real phenomenon that already has a satisfactory scientific explanation as if it were mysterious, or misrepresenting a scientifically explained phenomenon to make it mysterious, or even inventing phenomena. For example, the Bermuda Triangle "mystery" was a manufactured "mystery", i.e., it appeared to be a mystery only because facts about the circumstances, location, and the frequency of the alleged disappearances with respect to those in other parts of the oceans or seas were grossly misrepresented and even invented by writers promoting the alleged mystery.
26. Does not deliberately create mystery where none exists, by omitting crucial information or important details (Coker, 2001, p. 3). The Bermuda Triangle "mystery" (see property 25) is, again, a good example.
27. Does not actively look for mysteries as such (although it does look for potential falsifiers, falsifications, problems or puzzles, and anomalies, even recalcitrant ones) (Radner & Radner, 1982, pp. 32–4; Casti, 1990, pp. 57–8).
28. Does not ascribe a positive value to mystery or unexplainedness as such and in itself (ibid.; Radner & Radner, 1982, p. 34).

2.2.1.2 Non-specific Formal Properties of Propositions, Hypotheses, Theories, and Arguments

29. Internal logical consistency (non-self-contradiction). A scientific hypothesis or theory, at least in its finished form, must be self-consistent (Kuhn, 1977, pp. 321–2; Vollmer, 1993; Coker, 2001, pp. 2–3; Pigliucci, 2013, p. 24; Mahner, 2013, p. 38).
30. Is based on "the logic of inconclusive justification" (Kitcher, 1982, p. 35).
31. Generates (or has generated) hypotheses and theories through deductive and non-deductive processes (Bunge, 1960, pp. 6–10; Lack & Rousseau, 2016, p. 31).
32. Inducts from large and representative pools of observations or samples (which is required for increased randomisation). For generalisation purposes, it is better that the observations be made under a wide range of different conditions, or that the samples have been exposed to a range of different conditions (ibid., pp. 28, 45).
33. Avoids questionable arguments by elimination (Carey, 2011, pp. 112–113): "An argument by elimination is fallacious when it ignores other likely possibilities" (ibid., p. 113).

2.2.1.3 Non-specific Properties Related to the Contents of Problems, Propositions, Hypotheses, and Theories (Semantical Aspects, Metaphysical Aspects, Aboutness, etc.)

34. Has a particular domain or universe of discourse (Bunge, 2003, p. 260).
35. Its discourse and the terms it uses in it have meaning (cf. Lilienfeld et al., 2012, p. 27). It avoids using empty jargon (Carey, 2011, pp. 119–20).
36. Terms used have precise and unambiguous definitions (Coker, 2001, p. 4).
37. Presupposes philosophical background assumptions (Tuomela, 1985, p. 217), at least implicitly.
38. Its general outlook or philosophical background includes the epistemological principle that the world can be known objectively, at least partially and gradually (Bunge, 2003, p. 260).
39. Its general philosophical views contain a view of science as organised activity which aims at least in part at factually truthlike descriptions (Tuomela, 1985, pp. 216–8). This claim is independent of the debate between scientific realism, on the one hand, and scientific non-realism and scientific antirealism, on the other, because scientific antirealists admit that science can describe observables and aims to do it truthfully.
40. Its problematics consists exclusively of cognitive problems (actual or potential) that can be investigated (ibid.; Bunge, 2003, p. 260). Scientific theories account for real data and patterns (Dutch, 1982, p. 7): science does not deal with demonstrably spurious or invented pseudoproblems (ibid.; Mahner, 2013, pp. 39, 42). In Kuhnian terms, science has the capability to permit puzzle formulation and solution (Kuhn, 1996, p. 185).
41. Has a fund of knowledge which is a collection of up-to-date, testworthy and testable (and at least in part tested and confirmed) (yet corrigible) data, hypotheses, and theories which are compatible with its general philosophical views, as well as of other information previously incorporated, and of reasonably effective research methods, obtained by members of the research community at previous times (Tuomela, 1985, pp. 216–8; Bunge, 2003, p. 260). The possession of in-principle testability (property 51) is necessary for the full possession of this property.
42. Offers accounts (i.e., corroborated descriptions, corroborated explanations, retrodictions, analyses, models, etc.) (Hempel, 1942, 1965; Hempel & Oppenheim, 1948; Kitcher, 1982, pp. 30–54). It has the capability to be used to explain (Kuhn, 1977, p. 336). It has power to explain (Vollmer, 1993; cf. Mahner, 2013, p. 39) observable phenomena (Hempel, 1951, p. 74), especially intended data (Dutch, 1982, p. 7), and usually explains all the important data (ibid., pp. 7–8). It is better if scientific theories have broad explanatory power (Thagard, 2011, pp. 26–35). Consequences of scientific theories are themselves explanatory (Dutch, 1982, pp. 7–9). Science has explained new classes of facts (Thagard, 1988, p. 169). The possession of this property is necessary for the full possession of properties 46, 58, 82, 84, and 105.

43. Its propositions, hypotheses, or theories, taken individually, do not provide an in-advance universal explanation for any conceivable observation in a given domain (Derksen, 1993, pp. 33–6).
44. Postdicts (Hempel, 1942, 1965; Hempel & Oppenheim, 1948). It features postdictive power.
45. Predictive power: Science predicts (Hempel, 1942, 1965; Hempel & Oppenheim, 1948; Bunge, 1960, pp. 20–1). It features predictive power (cf. Mahner, 2013, p. 39) in regard to observable phenomena (Hempel, 1951, p. 74; Gruenberger, 1964, p. 1414).
46. Features proper, comprehensive (consilient) yet sufficiently specific theories (Tuomela, 1985, p. 235; Thagard, 1988, pp. 161, 169–70), or at least "models" (Tuomela, 1985, p. 232). In the first case, a science features "an independently testable framework of theory capable of supporting, connecting, and hence explaining" its claims (Grove, 1985, p. 237). Scientific theories have broad scopes (Kuhn, 1977, p. 322; Schick & Vaughn, 1995, pp. 177–8). The possession of property 42 is necessary for the possession of this property.
47. Scientific knowledge (except for that of non-general CFs such as history) is general (Russell, 1949, p. 10; Bunge, 1960, p. 17). Because of that, science has the capability to bring "order to phenomena that [...] would [otherwise] be individually isolated and, as a set, confused" (Kuhn, 1977, p. 322).
48. Explains using mechanisms (Thagard, 2011, p. 27). It does not claim to fully explain, predict, postdict, or retrodict by solely providing a mere dynamic scenario (a "story or plot" of how states of affairs develop/ed or will develop) without supporting causal mechanisms (and their supporting laws) or general principles—be this implicitly or explicitly—. That is, it indicates or implies explanatory connections in putative explanations (Radner & Radner, 1982, pp. 45–7; Casti, 1990, pp. 58–9; Coker, 2001, p. 5; Lack & Rousseau, 2016, p. 35).
49. Communicability: Scientific knowledge is not ineffable, but expressible. Scientific language can communicate information to whoever has been successfully trained to understand it (Bunge, 1960, pp. 14–5). Generally speaking, scientists, individually or communitarily, have the ability to, and know how to, communicate effectively to such trained individuals (Gruenberger, 1964, p. 1415; cf. Radner & Radner, 1982, p. 48). The more the receiving end is not well versed in the relevant subject matter or in the general scientific modes of procedure, the more this effectiveness can be impaired.
50. Usefulness, basically for a) harbouring conceptions that correspond to facts, and for b) developing technology (Bunge, 1960, pp. 22–3).
51. Testability, i.e., "the property of a proposition or theory of being able to be put to a test intended to either confirm or infirm it" (Bunge, 2003, p. 291). Science makes assertions than can be checked (Radner & Radner, 1982, p. 92) and avoids untestable explanations and predictions (Carey, 2011, pp. 117–9). The nature of testability depends on the field of inquiry (Tuomela, 1985, p. 213). An empirical scientific hypothesis or theory must be testable, at least in principle (Popper, 1959/2002, pp. 3, 10–6, 18; Tuomela, 1985, p. 213; Vollmer, 1993;

Mahner, 2013, p. 39; Boudry, 2013b, p. 86; Nickles, 2013, p. 103; Schick & Vaughn, 1995, pp. 172–4). The possession of this property is necessary for the full possession of many other properties in this list, e.g., properties 41, 52, 79, 96, 103–104, 112, 114, 122, 147, 187, and 210.

52. Its propositions are confirmable and disconfirmable, at least in principle (Bunge, 1960, pp. 6–9). The full possession of this property requires the possession of properties 51 (in-principle testability) and 54 (theoretical in-principle falsifiability).
53. Features theoretical boundary conditions (Lilienfeld et al., 2012, pp. 21, 28; Lilienfeld et al., 2015, p. 10).
54. Empirical science's theoretical in-principle falsifiability, i.e., its capability "of conflicting with possible, or conceivable observations" (Popper, 1959/2002, p. 11 and 1962, pp. 37–9; Radner & Radner, 1982, pp. 38–43; Casti, 1990, p. 58; Lilienfeld et al., 2012, pp. 21–3; Lack & Rousseau, 2016, pp. 30–1, 35). Empirical science makes assertions that can be refuted (Radner & Radner, 1982, p. 92); it is constructed in such a way as to be sensitive to possible counter-evidence (Grove, 1985, pp. 238–9). If a "theory is false, then it must be possible to show that it is false" (Kitcher, 1982, p. 37). The possession of this property is necessary for the possession of properties 7, 13, and 52.
55. Transparency/publicity: Scientific data, findings, methods, knowledge, etc. are public and open for all to scrutinise (Bunge, 1960, p. 14; cf. Lack & Rousseau, 2016, p. 40). "Scientific research processes must at least in principle be public throughout" (Tuomela, 1985, p. 212).
56. In-principle repeatability/replicability (cf. Mahner, 2013, p. 39). E.g., observations and scientific experiments must be reproducible by others with the same outcome (Tuomela, 1985, p. 212; Hansson, 1983, cited in Hansson, 2017; Vollmer, 1993). "Experiments must be precisely described so that they can be duplicated exactly or improved upon." (Coker, 2001, pp. 2, 6). The possession of this property is necessary for the possession of property 57.
57. The experiences (i.e., thinking processes, observations, or experiments) on which it is based have been repeated/replicated (cf. Lack & Rousseau, 2016, p. 41; Lilienfeld, 2016, pp. 268, 270). The possession of this property requires the possession of property 56.
58. High theoretical simplicity with regards to the isolated application of auxiliary hypotheses: "The explanation of facts F by a theory T [sometimes] requires [...] a set of auxiliary hypotheses A". These are assumptions "not independently acceptable" at the time and "not used to explain" any [other] phenomena (so that theirs are isolated applications, and therefore their "use is much more limited" than that of a theory). "Simplicity is a function of the size and nature of the set A needed by a theory T to explain facts F." (Thagard, 1988, pp. 83 (definition), 161, 170 (advocacy)). The possession of property 42 is necessary for the possession of this property.
59. Connectivity (Lilienfeld et al., 2012, pp. 21, 25–6; cf. Mahner, 2013, p. 39; Hoyningen-Huene, 2013, pp. 113–24; Lilienfeld et al., 2015, pp. 8–9; Mukerji & Ernst, 2022): A science is "generally a doctrine or body of doctrines" or

practices not "isolated and distinct from the science of its time" (cf. Tuomela, 1985, p. 229).

60. Unification. "A thriving science is not a gerrymandered patchwork but a coherent whole [...] setting the phenomena within a unified framework" (Kitcher, 1982, pp. 30–54).
61. Alleged phenomena are not at the very limit of detection (Langmuir ([1953] 1989, p. 44; Park, 2003).
62. Epistemic limitedness: Science is not omniscient, nor its theoreticians or practitioners claim that it is so (Lack & Rousseau, 2016, p. 47).
63. Fallibility/provisionality/tentativity in empirical knowledge (strictly speaking, there are "no [absolutely final] conclusive reasons for accepting a claim as true") (Bunge, 1960, pp. 9–10; Kitcher, 1982, pp. 32–3). Empirical science "cannot guarantee the truth of a conclusion, the best we can do is to reach conclusions as responsibly as possible, and accept that following (... [appropriate methodics]) vastly increases the likelihood of those conclusions being true". Empirical science is not "about final—or ultimate—answers", but "about a way of evaluating claims and gathering the evidence" that supports those claims (Lack & Rousseau, 2016, pp. 22–3, 31, 47). Empirical "scientific knowledge is [...] provisional" (Lilienfeld, 2016, p. 271). Its knowledge is only probable because confirmation is or might be incomplete, and because of that, it is to be deemed provisional. There is no final confirmation (Bunge, 1960, pp. 6–10). Its theoreticians or practitioners do not claim all this to be otherwise, unless they are using shorthand, or they are suspending doubt for methodological or pedagogical reasons. It is important to notice technical language in this respect, e.g., the expression "scientific fact" is commonly used to mean "sufficiently established hypothesis that we believe to be, or treat as if it were, true", allowing that it could be false just in case the relevant scientific consensus was wrong.
64. Scientific accounts are intersubjectively valid, i.e., have publicly accessible supporting evidence (Nagel, 1961, p. 449). Scientific studies "result in publicly verifiable knowledge" (Gruenberger, 1964, p. 1414; Lawson, 2007, p. 4).

2.2.1.4 Non-specific Properties Related to Epistemic Attitudes and Values (Axiological Properties), Individual or Social

65. Rationality (cf. Mahner, 2013, p. 39), i.e., conformity of beliefs with reasons to believe and of actions with reasons for action. "Scientific activity is both end-rational and means-rational" when properly pursued. Its means-end rationality is characterised by its objectivity (our property 2), criticalness (our property 118), progress (our property 84) (Tuomela, 1985, pp. 212–8), and all the rest of its properties that are listed here. Concepts, judgements, and arguments feature logical rules, and are organised in systems, not chaotically (Bunge, 1960, pp. 6–17; Simon, 1969, p. 414; cf. Bunge, 2013, pp. 21–2).
66. Does not attack logic; accepts it (Dutch, 1982, pp. 11–2; Mahner, 2013, p. 38).
67. Features curiosity (cf. Bunge, 2013, p. 26).

68. Is a creative activity (Tuomela, 1985, p. 217).
69. Looks for objective knowledge/truths (cf. Bunge, 2013, pp. 21–2).
70. Strives for knowledge of causal mechanisms of reality (ibid., p. 13).
71. Recognises the limitations of speculative theorising (Radner & Radner, 1982, p. 91).
72. Recognises the limitations of sketchy theorising (ibid.).

Social

73. The community of theorists is concerned with reducing the number of unexplained phenomena (cf. Kitcher, 1982, p. 49).
74. Theoreticians care about evaluating their theories in relation to alternative theories (Thagard, 1988, pp. 166–71; 2011, p. 27).
75. The community proposing the theory changes its mind when the evidence compels it to do so (Lilienfeld et al., 2012, p. 12).
76. The community of theorists is concerned with generating correct predictions (cf. Kitcher, 1982, p. 49).
77. The community of theorists do not forswear "excursions into new areas of investigation where embarrassing questions might arise" (ibid.).
78. Features a low degree of unjustified selectivity in the community of practitioners and theorists when considering factors, collecting data, applying the laws of their theory, and considering confirmations and disconfirmations (Thagard, 1978, pp. 227–8; Hansson, 2013, pp. 72–3; Boudry, 2013b, p. 91; Shermer, 2013, p. 217; Ruse, 2013, p. 226).
79. Scientists show *regard of refuting information*: observations or experiments that conflict with a theory are not neglected (Hansson, 1983, cited in Hansson, 2017). The possession of properties 51 and 52 is necessary for the possession of this property.
80. Avoids using spurious strategies to make disconfirmations appear as confirmations (Boudry, 2013b, pp. 87, 89, 95); or, more generally, avoids distorting the data that are unfavourable to the theory (Bunge, 1984, p. 41).

2.2.1.5 Non-specific Epistemic Properties of Practices, Propositions, Hypotheses, and Theories, and of the Evolution of the Discipline

81. Parsimony of / Unification of problem-solving strategies: "Good theories consist of just one problem-solving strategy, or a small family of problem-solving strategies, that can be applied to a wide range of problems. The theory succeeds as it is able to encompass more and more problem areas. Failure looms when the basic problem-solving strategy (or strategies) can resolve almost none of the problems in its intended domain without the "aid" of untestable auxiliary hypotheses". Success is achieved through a *unified* theory. "Problems are solved by invoking the same pattern of reasoning, or *problem-solving strategy*, again

and again". "*Unification* is the result of applying a small family of problem-solving strategies to a broad class of cases." (Kitcher, 1982, pp. 30–54).

82. But because all human science has been incomplete so far, an important property of science is fecundity/fruitfulness. Fruitfulness (Gruenberger, 1964, p. 1414; cf. Mahner, 2013, p. 39; Schick & Vaughn, 1995, pp. 174–7) of research findings of "new phenomena or previously unnoted relationships among those already known" (Kuhn, 1977, p. 322). It "grows out of incompleteness" when a theory raises questions and "opens up new and profitable lines of investigation" (i.e., areas of research). "A great scientific theory [...] opens up new areas of research". "Typically, a flourishing science is incomplete. At any time, it raises more questions than it can currently answer. But incompleteness is no vice. On the contrary, incompleteness is the mother of fecundity. Unresolved problems present challenges that enable a theory to flower in unanticipated ways. They also make the theory hostage to future developments. A good theory should be productive; it should raise new questions." (Kitcher, 1982, pp. 30–54; Vollmer, 1993). General fecundity/fruitfulness encompasses the four following subproperties, which are ways of solving problems—providing answers—: i) Fecundity/fruitfulness of corroborated descriptions (Kitcher, 1982, pp. 30–54); ii) Fecundity/fruitfulness of (preferably unifying, preferably corroborated) explanations (ibid.); iii) Fecundity/fruitfulness of predictions (ibid.); and iv) Fecundity/fruitfulness of successful predictions. This is "one important way" in which a theory can be justified and "win our acceptance" (ibid.). The possession of property 42 is necessary for the full possession of this property.
83. Promise: Promises further fruitful development (ibid., pp. 48, 54).
84. Progressiveness: Progress is, or has been, made (Tuomela, 1985, pp. 212–8; Grove, 1985, pp. 237–8; Thagard, 1988, p. 168; 2011, pp. 27–31; cf. Mahner, 2013, p. 39). "As time goes on, more and more is learned about the [...] processes under study." (Coker, 2001, p. 6). Progresses over time: develops new theories that explain newly discovered facts or even new facts (Thagard, 1988, pp. 168–70; 2011, pp. 27–31). New information is uncovered, and discoveries are made. New theories are proposed, and old concepts are modified or discarded in light of new discoveries (Coker, 2001, pp. 3, 6). Progressiveness as opposed to immutability/staticity in the face of problems (taking into account what alternative explanations are available) (Thagard, 1988, pp. 168–9). In Lakatosian terms, the scientific research programme in question has undergone/is undergoing theoretically progressive problemshift/s (Lakatos, 1970, p. 118). The possession of property 42 is necessary for the full possession of this property.

2.2.2 *Properties that Are Specific to SBS*

2.2.2.1 Specific Properties Related to Practices, Methods, and Techniques

85. Bold hypothesising (cf. Bunge, 2013, p. 27).
86. Originality against common sense (cf. Bunge, 2013, p. 48).
87. General meticulosity (cf. Bunge, 2013, pp. 27–48).
88. General rigorosity (cf. Lilienfeld, 2016, p. 268) in thinking, methodics, representation, etc. "[S]cience is [...] demarcated from non-science [...] by [...] requirements governing representational rigour with respect to both theoretical claims and accounts of observations and experiments" (Ladyman et al., 2007, p. 28).
89. General precision: Features a high degree of precision of formulation (Hempel, 1951, p. 74; Bunge, 1960, pp. 13–4; Lilienfeld et al., 2012, p. 27; cf. Bunge, 2013, pp. 27–48).
90. Does not exaggerate (Lilienfeld, 2016, p. 272). It is cautious, not hyperbolic or sensationalistic (Lack & Rousseau, 2016, pp. 19, 42–3).
91. General concision (cf. Bunge, 2013, pp. 27–48).
92. General clarity; science is not vague (Glymour & Stalker, 1990, p. 94) or ambiguous. It features a high degree of clarity of formulation (Hempel, 1951, p. 74). It purifies notions, defines concepts, creates language(s), measures, and records (Bunge, 1960, pp. 13–4; cf. Bunge, 2013, pp. 27–48). Clear conceptual sense is made of phenomena (cf. Tuomela, 1985, p. 232).
93. Does not use obscurantist language (Lilienfeld et al., 2015, pp. 9–10).
94. Is "painstaking" (Simon, 1969, p. 414).
95. Systematically gathers information (Hoyningen-Huene, 2013; Lack & Rousseau, 2016, p. 31).
96. Uses, and abides by, evidence (Mahner, 2013, p. 39), i.e., empirical data relevant to a hypothesis or theory. This data is obtained through carefully controlled, demanding, and precise observation/study or, preferably when possible, demanding, randomised controlled experiments (Simon, 1969, p. 414) or rigorous trials (Gruenberger, 1964, p. 1414; Lawson, 2007, p. 4; Bunge, 2013, pp. 1–48). It is better if this evidence is compelling (Lilienfeld, 2016, p. 268). Science devises rigorous tests for its propositions (Dutch, 1982, p. 12). It "convinces by appeal to the evidence, [and] by arguments based upon logical and/or mathematical reasoning" (Coker, 2001, p. 6). The possession of in-principle testability (property 51) is necessary for the full possession of this property.
97. Does not use handpicked ("cherry-picked") data, evidence, examples, or outcomes (Lack & Rousseau, 2016, pp. 28, 45–6; Lilienfeld, 2016, p. 272; Mukerji & Ernst, 2022) if "they are not representative of the general category that the investigation refers to." (Hansson, 1983, cited in Hansson, 2017).
98. Does not suppress data that clashes with its preconceptions (Coker, 2001, p. 7).

99. Factual screening and sorting: Features screening and sorting of putative facts that is appropriate to separate pseudofacts from facts (Radner & Radner, 1982, p. 70), and a lack of reluctance to weed out pseudofactual, suspicious, or low-quality putative evidence (ibid., p. 38). Does not appeal to widespread misunderstandings (Coker, 2001, p. 1).
100. Is evidenced: Presents coherent evidence/proof of its propositions (Dutch, 1982, p. 12; cf. Tuomela, 1985, p. 232). There is no lack of evidence (Mukerji & Ernst, 2022; cf. Coker, 2001, p. 4). It does not falsely claim to have provided proof elsewhere, or avoids giving proof citing didactical reasons (Derksen, 2001, pp. 340–1). It incorporates sufficient untainted evidence (Derksen, 1993, pp. 21–3).
101. Understands where the burden of proof lies (Dutch, 1982, p. 12). Does not reverse the burden of proof (Radner & Radner, 1982, p. 82; Pigliucci & Boudry, 2014; Lilienfeld et al., 2015, p. 8). Applies the methodological principle that those who make a claim should bear the burden of its proof (cf. Bunge, 2013, pp. 21–2).
102. Uses mainly careful scrutiny (Lilienfeld, 2016, pp. 268–9, 270).
103. Its methodics (usually misnamed "methodology") consists exclusively of scrutable (checkable, analysable), criticisable, justifiable, explainable, testable and historically correctable procedures which are appropriate scientific methods (scientific methods are safeguards against error (Lilienfeld et al., 2012, p. 14)) and techniques, in the first place the general scientific method (Tuomela, 1985, pp. 216–8; Bunge, 2003, p. 260). The possession of in-principle testability (property 51) is necessary for the full possession of this property.
104. Methodical scientifcity: Has adopted the general scientific method as its main strategy for dealing with its cognitive problematics (Bunge, 2003, p. 260; 2013, p. 48). One definition of the general scientific method is "[t]he sequence: Survey of a body of knowledge → Choice of problem in this body of knowledge → Problem formulation or reformulation → Application or invention [and application] of an approach for handling the problem → Tentative solution (hypothesis, theory, experimental design, measuring instrument, etc.) → Checking the tentative solution → Evaluating the tentative solution in the light of both test[/s] and background knowledge → Revision or repetition of any of the previous steps → Estimation of impact on background knowledge → Final evaluation (until new notice)" (Bunge, 2003, p. 180). "The method of science is not an immutable given but a plastic collection of ideas and rules" (Tuomela, 1985, p. 210). It changes or has changed, however slowly, as a result of scientific inquiry (formal or empirical) (Bunge, 2003, p. 260). The reliability of all the methods and techniques of science can be independently tested (cf. Mahner, 2013, p. 39). "The scientific method is self-correcting" (Schick & Vaughn, 1995, p. 159). The possession of in-principle testability (property 51) is necessary for the full possession of this property.

105. Independent recognisability of correctness: Its descriptions and explanations can be independently recognised as correct (Kitcher, 1982, p. 49). The possession of property 42 is necessary for the possession of this property.
106. Avoids the idea of having the correct explanation by default (Dutch, 1982, p. 11).
107. Is analytical but also generalises. It recognises that syntheses cannot be cognitively apprehended without analyses, and operates consequently (Bunge, 1960, pp. 12–3).
108. Avoids *belief in personal authority* other than the one derived from knowledge, training, experience, study, and reflection (Hansson, 1983, cited in Hansson, 2017). It recognises this personal intellectual authority, but, at the same time, has a non-authoritarian epistemology, i.e., it does not accept epistemic justification on the basis of that authority (Gruenberger, 1964, pp. 1414–5; cf. Tuomela, 1985, pp. 227–8).
109. Recognises the epistemic authority of science or of the scientific community (Dutch, 1982, p. 12). Does not unjustifiedly distrust experts or their authority (Shelton, 2004, p. 174).
110. Does not claim any special insight for a special "initiated" group other than the insight coming from knowledge, training, experience, study, and reflection (Derksen, 1993, pp. 32–33).
111. Does planned (and mostly non-sloppy, methodical) research and abides by it (Bunge, 1960, pp. 16–7; Coker, 2001, p. 1; Bunge, 2013, pp. 1–48). Features theoretically-directed investigation, not just trial and error. Observations and experiments are better if designed in the light of (plausible) hypotheses (cf. Bunge, 2013, pp. 1–48).
112. Checks for variation of variables. Experimentation is epistemically better than passive observation because experiments involve "the control of variables and therefore the possibility of finding and testing" hypotheses, including causal hypotheses (Bunge, 2013, pp. 1–48; Lack & Rousseau, 2016, p. 45). The possession of in-principle testability (property 51) is necessary for the full possession of this property.
113. Uses control groups (ibid., pp. 46–7).
114. Uses blind (masked) testing. It is better if the blinding is double, and even better if it is triple (cf. ibid.; Lilienfeld, 2016, p. 270). The possession of in-principle testability (property 51) is necessary for the full possession of this property.
115. Gets independent confirmation of the alleged facts (Shelton, 2004, p. 174).
116. Systemicity: When possible, science treats reality systemically, i.e., like a system that can be conceptually dismantled and analysed. "[S]ystemicity implies analyticity, for understanding a system involves analysing it. Systemicity also includes the valid component of holism, namely the thesis that 'a whole is more than the set of its parts'" (Bunge, 2013, pp. 13–5).
117. Avoids dogmatism ("the refusal to face facts and debate rationally") and dogmatic assertions (Thagard, 2011, pp. 27–8; definition: Bunge, 2013, p. 24; cf.

Lack & Rousseau, 2016, pp. 20, 23). This is advisable because of property 63 (the general fallibility/provisionality/tentativity of empirical science).

118. Criticalness: Science incorporates a critical, sceptical and examinatory attitude (Tuomela, 1985, p. 212; Mahner, 2013, p. 39). Science is critical (Derksen, 1993, pp. 36–7). It uses constructive criticism within a philosophical matrix (cf. Bunge, 2013, p. 32). Scientific scepticism: "a mindset of doubt coupled with an open-mindedness to new claims." "[W]e must strike a healthy balance", a "judicious mix", "between doubt and open-mindedness: this balance is the approach often termed [scientific] scepticism" (Lilienfeld et al., 2012, pp. 12–13; Lilienfeld et al., 2015, pp. 4–5). Science features a certain scepticism "even towards one's most cherished theories. Blind commitment to a theory is [...] an intellectual crime." (Lakatos, 1978, p. 1). Science is self-critical (cf. Derksen, 2001, pp. 332–7; Lilienfeld, 2016, p. 272); it "relies on – and insists on – self-questioning" (Coker, 2001, p. 6).
119. Welcomes criticism (cf. Tuomela, 1985, p. 228) instead of resisting it (Grove, 1985, pp. 239–40). Lends itself to criticism and counterarguments (Radner & Radner, 1982, p. 92). Considers opposing views (ibid., pp. 95–7).
120. Revises or checks, or at least tends to revise or check, in light of new evidence or criticism. Satisfactorily accounts for new evidence or answers original criticism (Radner & Radner, 1982, pp. 51, 95; Coker, 2001, p. 1; cf. Derksen, 2001, pp. 332–7) in order to strengthen the scientific content of its proposals, or at least tends to (Radner & Radner, 1982, p. 96).
121. Avoids emotional attachment to a hypothesis or theory (Shelton, 2004, p. 175).
122. Is "continually striving to find ways of subjecting" its "most cherished beliefs to rigorous testing" (Lilienfeld et al., 2012, p. 12). Tests and analytical thinking make it difficult for science to fool itself or to avoid facing facts (Coker, 2001, p. 6). The possession of in-principle testability (property 51) is necessary for the full possession of this property.
123. Is quick to correct mistakes (Coker, 2001, p. 7).
124. As confirmation is easier to accomplish than falsifiability, science does not give excessive weight to confirmation over falsifiability or over falsification (Lilienfeld et al., 2012, pp. 21, 23–4; Lilienfeld et al., 2015, p. 8; cf. Lack & Rousseau, 2016, p. 39).
125. The doctrine does not apply its problem-solving strategies with auxiliary hypotheses that can be "tested" only in their applications (Kitcher, 1982, p. 48).
126. Avoids sectoral or sectional thinking that "a) detaches and isolates things, properties, and processes that in fact go together; b) tends to "anchor" in first impressions, data, or conjectures (see Kahneman, 2011)"; and c) balkanises "instead of building interdisciplinary bridges" (cf. Bunge, 2013, p. 46).
127. Does not pitch claims directly to the popular mass media first (Park, 2003).
128. Is scientifically peer-reviewed (Park, 2003; Lilienfeld, 2016, pp. 268–9): Its "findings are expressed primarily through scientific journals that are [scientifically] peer-reviewed and maintain rigorous standards for honesty and accuracy" (Coker, 2001, p. 6). "[S]cience is [...] demarcated from non-science [...] by institutional norms: requirements for rigorous peer review before

claims may be deposited in 'serious' registers of scientific belief' (Ladyman et al., 2007, p. 28).

129. Avoids using recognised formal or informal fallacies (including affirming the consequent, denying the antecedent, *argumentum ad baculum*, *argumenta ad hominem*, *argumenta ad ignorantiam* (and avoids exploiting uncertainty), *argumentum ad misericordiam*, *argumenta ad populum*, *argumentum ad verecundiam*, fallacies of accident, *post hoc ergo propter hoc*, fallacious *petitio principii*, use of equivocation or amphibology, fallacy of accent, fallacy of composition, fallacy of division, and many others) (Coker, 2001, p. 3; Carey, 2011, pp. 113–5, 121–2). Avoids "straw man" arguments (Dutch, 1982, p. 11).
130. Does not argue from errors or probable errors, or from unverified suspect claims, without checking on them (Coker, 2001, pp. 3–4).
131. Does not argue from irrelevancies (e.g.: "Scientists don't know everything!") (Coker, 2001, p. 4).
132. Does not excessively rely on anecdotes, hearsay, eyewitness testimony, rumour, or news reports (Coker, 2001, p. 1; Park, 2003; Carey, 2011, p. 112; Lilienfeld et al., 2012, pp. 21, 24–5; Lilienfeld et al., 2015, p. 9; Lilienfeld, 2016, p. 272).
133. Does not consider that anecdotes trump (other kinds of) evidence (cf. Lack & Rousseau, 2016, p. 39). Does not use the "Piling Up The Anecdotes" strategy: focusing "almost entirely on recording the 'hits' - of cases where the method was applied and" success achieved. "This approach [...] is deemed 'scientific' because, in the minds of [...] followers, a huge amount of data has been built up in support of their theory." (Tuomela, 1985, p. 228; quotes: Law, 2011, Chap. 7; Law, 2019, p. 110).
134. Does not have a "grab-bag" approach to evidence, i.e., "the attitude that sheer quantity of evidence makes up for any deficiency in the quality" of individual pieces of evidence (Radner & Radner, 1982, pp. 36–8; Casti, 1990, p. 58).
135. "Does not advocate or market unproven practices or products." (Coker, 2001, p. 6).
136. Does not argue from spurious similarity to science (Radner & Radner, 1982, pp. 43–5; Casti, 1990, p. 58).
137. Does not do "research by exegesis" (Radner & Radner, 1982, pp. 47–51) or by "literary interpretation" (Casti, 1990, p. 59).
138. Does not use viciously circular reasoning (Radner & Radner, 1982, p. 97; Vollmer, 1993; Mahner, 2013, p. 38).
139. Does not use metaphors (although it can use analogy) (Radner & Radner, 1982, p. 97).
140. Does not oversimplify (ibid., p. 92; Lack & Rousseau, 2016, pp. 42–3).
141. Examines its own philosophical presuppositions (cf. Bunge, 2013, p. 43).
142. Does not aprioriy or otherwise improperly restrict acceptance of data interpretations to certain particular interpretations (Derksen, 1993, p. 23).
143. Resists the allure of seeing sense in spectacular coincidences (advocacy: Derksen, 1993, pp. 26–30; definition: Derksen, 2001, p. 346).
144. Does not feature excessive pretension (Derksen, 1993, pp. 36–7).

145. Does not use striking but irrelevant examples (Derksen, 2001, pp. 338–40).
146. Does not use the strategy of the favourable compromise (ibid., pp. 341–3). Some of the varieties of this strategy are: i) making claims that are outside what is epistemically justified and then offering to trade off by renouncing some epistemically unjustified claims in favour of maintaining other epistemically unjustified claims; ii) making claims that are outside what is epistemically justified and then apparently compromising by accepting more moderate claims that are closer to, but still outside, what is epistemically justified.
147. Avoidance of cognitive errors and biases: Tends to, tries to, and mostly succeeds in its attempt to, avoid all the known cognitive errors and biases, including: naïve realism, self-confirmation bias, premature closure, belief perseverance, illusory correlation, hindsight bias, groupthink, overreliance on heuristics, availability heuristic, anchoring heuristic, affect heuristic, representativeness heuristic, base-rate neglect, bias blind spot, unpacking, aggregate bias, search satisfying, diagnostic overshadowing, diagnosis momentum, psych-out error, pathology bias, etc. (Lilienfeld et al., 2012, pp. 14–21; cf. Derksen, 2001, pp. 337–8; Lack & Rousseau, 2016, p. 21). This is served by implementing controlled tests (ibid., p. 48) (which requires the possession of in-principle testability (property 51)):
148. Uses finely honed and systematic safeguards to compensate for cognitive biases (Lilienfeld, 2016, pp. 270, 272).
149. Does not change its claims and denies that it did so (Glymour & Stalker, 1990, p. 94).
150. Does not claim that it is necessary to believe in order to understand (ibid.).
151. Does not begin with a hypothesis and then look only for items which appear to support it (Coker, 2001, p. 1).
152. Investigates or tests alternative possibilities (Coker, 2001, p. 1; Schick & Vaughn, 1995, pp. 224–5).
153. Tenable explanations are not abandoned without being replaced, in a way in which the new theory leaves much more unexplained than the previous one (Hansson, 1983, cited in Hansson, 2017).
154. Concern for evidentiary criteria: Is not indifferent to criteria of valid evidence (e.g., that evidence must be obtained through meaningful, controlled, repeatable observations or experiments) (Coker, 2001, p. 2).
155. Does not rely on emotional impact (Park, 2003). Does not attempt to persuade with rhetoric or propaganda (Coker, 2001, p. 3). Does not appeal to emotion, sentiment, or distrust of established facts (ibid., p. 4). Does not ignore, misinterpret, or mischaracterise scientific literature, or evidence (ibid., p. 2; Mukerji & Ernst, 2022). Does not attempt to persuade with ignorance or misrepresentation (Coker, 2001, p. 3).
156. Does not rely heavily on subjective validation (ibid., p. 2).
157. Does not depend on "arbitrary conventions of human culture" (ibid.).
158. Does follow up (ibid.).
159. Checks its sources (ibid., p. 1).
160. Deems tradition deserving to be studied but not worshipped (cf. Bunge, 2013, p. 26). Does not respect ideas in direct proportion to their antiquity (Coker,

2001, pp. 3, 5; Park, 2003). This is the *ad antequitem* fallacy (Lilienfeld et al., 2012, pp. 21, 27).

161. Avoids radicalism (cf. Bunge, 2013, p. 24).
162. The doctrine's problem-solving strategies are not an opportunistic collection of unmotivated and unrelated methods (Kitcher, 1982, p. 48).
163. The doctrine does not forswear excursions into new areas of investigation where embarrassing questions might arise (ibid., p. 49).
164. Ockham's razor: the principles that a) "entities are not to be multiplied without necessity" (also called the principle of parsimony), and that b) we should favour "the simpler of two hypotheses that account for the same data" as long as "neither of the rival hypotheses enjoys the support of additional [accepted] hypotheses" (Gruenberger, 1964, p. 1414; Vollmer, 1993; Bunge, 2003, pp. 200–1; Mahner, 2013, p. 39).
165. Does not use the "It Fits!" strategy: unjustifiedly making the theory "consistent with - be made to 'fit' - the available evidence" by sheer use of ingenuity or imagination in the composition of the theory itself (Law, 2011, Chap. 2; Law, 2019, p. 110).
166. Avoids the Fulton non sequitur/"Galileo fallacy"/"Galileo Gambit": alleging that a rejected idea should still be seriously considered because many rejected ideas were later recognised as great theories, discoveries, innovations, or inventions. The namesakes here are Robert Fulton (the developer of the first commercially successful steamboat) and Galileo Galilei. A stronger and wronger version of this fallacy takes the form of believing that the rejected idea must be correct (Gruenberger, 1964, p. 1415; Dutch, 1982, p. 10; Glymour & Stalker, 1990, p. 94; Coker, 2001, p. 3).
167. Does not presuppose ulterior motives motivating opponents (Dutch, 1982, p. 10; Grove, 1985, p. 239). Does not feature paranoia (Gruenberger, 1964, p. 1415; Grove, 1985, p. 239; Glymour & Stalker, 1990, p. 94; Park, 2003). Does not feature imaginary conspiracies (Coker, 2001, p. 4).
168. Is not overly speculative (Dutch, 1982, p. 9; cf. Glymour & Stalker, 1990, p. 93).
169. Unambiguity of the consequences of the methodology behind the theory (Dutch, 1982, p. 9).
170. Does not exaggerate uncertainty, differences of opinion among scientists, or normal experimental errors (ibid., p. 12). Avoids "fabricating the appearance of ongoing controversy [(…) when] scientists have achieved a consensus (Oreskes, 2004; Harker, 2017)." (Hirvonen & Karisto, 2022, p. 710). Does not confound assumptions made at different levels of certainty (Dutch, 1982, pp. 10–11).
171. Does not appeal to magical thinking and avoids it, e.g., the law of similarity/resemblance and the law of contact (cf. Tuomela, 1985, pp. 220–3; Thagard, 1988, p. 162; 2011, p. 27). "[R]esemblance thinking […] infers that things are causally related merely because they are similar." (ibid.). "Magic, sorcery, witchcraft […] are based on spurious similarity, false analogy, false cause-and-effect connections, etc. That is, inexplicable influences and connections

between things are assumed from the beginning—not found by investigation." (Coker, 2001, p. 5). Nonetheless, "not all reasoning involving similarity is resemblance thinking. The detection of similarities is a pervasive feature of thought and only becomes illegitimate when the leap is made from mere similarity to causal connection. Analogies and metaphors are very important in scientific and everyday reasoning, but analogical inference is not a form of resemblance thinking. [...] Analogy here functions primarily as a heuristic device, although it may also play a subsidiary role in the validation of the existence of a causal relation between phenomena." (Thagard, 1988, p. 165). But in this case analogies must be "supported":

> thinking about a puzzle in terms of something similar but better understood can lead to possible explanations. But [(...) there is a] need for independent testing of the explanation arrived at in this way. Analogies and similarities are fallaciously exploited when the fact that an explanation works in one case is given as evidence for the correctness of a similar explanation in another case. At the very most, a well-chosen similarity guides us to a possible explanation; it should not be thought to provide evidence that the explanation is correct. (Carey, 2011, p. 116)

172. Does not involve anachronistic thinking, i.e., thinking going back to old, refuted theories and assumptions (Casti, 1990, p. 57; cf. Tuomela, 1985, p. 228; cf. Coker, 2001, p. 5; cf. Mahner, 2013, p. 39), it does not take up a falsified and rejected theory again unless it is seen how to make it work where it did not work before (Radner & Radner, 1982, pp. 29–32). It does not cite obsolete research "or quote debunked experimental results" (ibid., pp. 96–7; quote: Mukerji & Ernst, 2022). "[W]hen an experiment or study has been shown to be questionable, it is [...] dropped from the list of confirming evidence" (Casti, 1990, p. 58).
173. Does not appeal to myths (Radner & Radner, 1982, pp. 34–6; Casti, 1990, p. 58; cf. Tuomela, 1985, p. 228) or religion. Does not base itself on "ancient religious or mythological works" (Coker, 2001, p. 1).

2.2.2.2 Criteria that Concern Specific Formal Properties of Propositions, Hypotheses, Theories, and Arguments

174. Has a formal background, which is a collection of up-to-date logical and mathematical theories (Simon, 1969, p. 414; Bunge, 2003, p. 260) which can be used, e.g., to sharpen theory formation and to process data (Tuomela, 1985, pp. 216–8). Science does not shun conceptually and logico-mathematically exact thinking (cf. ibid., p. 228). Science "convinces by appeal to the evidence, [and] by arguments based upon logical and/or mathematical reasoning" (Coker, 2001, p. 6).

2.2.2.3 Specific Properties Related to the Contents of Problems, Propositions, Hypotheses, and Theories (Semantical Aspects, Metaphysical Aspects, Aboutness, etc.)

175. Its general philosophical views contain an adequate view of the scientific method (Tuomela, 1985, pp. 216–8).
176. *Changes or has changed,* however slowly, typically *as a result of* research results obtained in and through *scientific inquiry* (formal or empirical), rather than, e.g., "as a result of [non-scientific] ideological or political pressures, or of 'negotiations' among researchers" (Tuomela, 1985, pp. 216–8, cf. 228; quote: Bunge, 2003, p. 260).
177. Does not use the "Explain THAT!" strategies: alleging that one is proposing the best explanation for alleged facts that are not established to have occurred; or otherwise unsoundly applying the argument to the best explanation. "Notice that anything we observe can be neatly and easily explained by positing a hidden agent with extraordinary powers and a desire to bring it about." (Law, 2019, p. 111).
178. Scientific knowledge is systematic (Bunge, 1960, p. 17; Hoyningen-Huene, 2013). Science "aims at the systematic and organized pursuit of [...] knowledge" (Tuomela, 1985, p. 211).
179. Features a high degree of specificity in the theoretical description of the causal mechanisms that its theories propose (Mahner, 2013, p. 38; Boudry, 2013b, pp. 86, 89; Goode, 2013, pp. 146–7, 156–7; Fales, 2013, p. 259). We have to "have [...] details of what that explanation" is (Carey, 2011, p. 116).[6]
180. Does not feature dubious forms of science-sounding "*mechanics* – positing various forces, entities, fields, stuffs, and/or channels, which can be blocked, enhanced, and so on" (Law, 2019, p. 112).
181. General legality: Scientific knowledge is legal (Bunge, 1960, p. 17). "Scientific investigation aims to disclose the general principles that govern the workings of the universe" (Kitcher, 1982, p. 35). "Science, as its name implies, is primarily knowledge; by convention it is knowledge of a certain kind, the kind, namely, which seeks general laws connecting a number of particular facts" (Russell, 1949, p. 10).
182. Explanatory legality: Science's explanations are proposed with respect to laws (Hempel, 1942, 1965; Hempel & Oppenheim, 1948; Bunge, 1960, pp. 19–20).
183. Empirical adequacy within an intended domain of application ("consequences deducible from a theory should be in demonstrable agreement with the results of existing experiments and observations" in its intended domain, and that

[6] However, Sagan (1976, p. 2) indicated a good point with respect to this criterion: "This is certainly [...] relevant [...] but by itself it's unconvincing. No mechanism was known for continental drift (now subsumed in plate tectonics) when it was proposed by Alfred Wegener in the first quarter of the twentieth century to explain a range of puzzling data in geology and palaeontology. [...] Objections to pseudoscience on the grounds of unavailable mechanism can be mistaken".

agreement should be "not only quantitative [...] but qualitative as well") (Kuhn, 1977, p. 321).

184. A high degree of theoretical confirmation (success of tests or of predictions) (Vollmer, 1993). Its theories have been confirmed by experiential evidence to a good extent (Hempel, 1951, p. 74; Glymour & Stalker, 1990, p. 94). That is: (a) a high degree of empirical adequacy of the theory to the data already known when the theory is formulated; and (b) a high degree of theoretical predictive success, that is, a high degree of empirical adequacy of the theory to the new data that is collected after the formulation of the theory (in connection with the virtue of explaining new facts) (Shermer, 2013, p. 208; Ladyman, 2013, p. 54). The doctrine's problem-solving strategies do not encounter recurrent difficulties in a significant range of cases of the type for which it was proposed (Kitcher, 1982, p. 48). Science is based on well-confirmed general theories (cf. Tuomela, 1985, p. 228).
185. Does not have *built-in subterfuge* (Hansson, 1983, cited in Hansson, 2017), i.e., a way of rendering the theory immune to critiques, falsifications, disconfirmations, etc., or to rescue it from them, that is part of the theory itself.
186. Independent testability of auxiliary hypotheses: it is possible to test a theory's auxiliary hypotheses independently of the particular cases for which they were introduced. "An auxiliary hypothesis ought to be testable independently" of the particular problem it was introduced to solve, "independently of the theory it is designed to save" (Kitcher, 1982, pp. 30–54). The possession of this property is necessary for the possession of a corresponding property of method: property 125.
187. Independent confirmation of auxiliary hypotheses: a theory's auxiliary hypotheses are confirmed independently of the particular cases for which they were introduced. An auxiliary hypothesis ought to be confirmed independently of the particular problem it is introduced to solve, independently of the theory it is designed to save (Kitcher, 1982, pp. 30–54). The possession of in-principle testability (property 51) is necessary for the possession of this property.
188. Simplicity (Kuhn, 1977, p. 322; Schick & Vaughn, 1995, pp. 178–80). It is formally simple (Hempel, 1951, p. 74). Uses simple theories (Thagard, 2011, p. 27).
189. Contiguity: For each particular research field, there is at least one other *contiguous* scientific research field (whose investigators are similarly capable of scientific inference, action and discussion as the members of the first scientific research field are) such that (a) the two fields share some items in their general outlooks, formal backgrounds, specific backgrounds, funds of knowledge, aims, and methodics; and (b) either the domain of one of the two fields is included in that of the other, or each member of the domain of one of the fields is a component of a system in the domain of the other (Tuomela, 1985, pp. 216–8; Bunge, 2003, p. 260).
190. Bulk external consistency, i.e., compatibility with the bulk of: "known facts", well-confirmed knowledge, and other theories contemporaneously accepted

(Kuhn, 1977, pp. 321–2; Dutch, 1982, p. 8; cf. Glymour & Stalker, 1990, p. 94; Vollmer, 1993; Schick & Vaughn, 1995, pp. 180–1; Coker, 2001, p. 4; Park, 2003; Bunge, 2013, p. 21; cf. Mahner, 2013, p. 39). The bulk of consequences of the theory are externally consistent (Dutch, 1982, pp. 8–9).[7]

191. Features a lack of, or small total amount of, or unimportance of, unsolved problems, including anomalies (Thagard, 1978, pp. 227–8).
192. Embedded in its theories there is a zero or low degree of arbitrary selectivity in applying principles, considering factors, and collecting data (Boudry, 2013b, p. 91).
193. The knowledge it provides is more systematic than other kinds of knowledge, in particular everyday knowledge, about the same subject matters (Hoyningen-Huene, 2013, pp. 14, 23). The possession of systematic knowledge (property 178) is necessary for the possession of this property.

2.2.2.4 Specific Properties Related to Epistemic Attitudes and Values (Axiological Properties), Individual or Social

194. Ethos: Its general outlook includes "the ethos of the free search for truth [...] and system" (Bunge, 2003, p. 260) (cf. properties 3, 39, 65, 69, 95, 116, 178, 195–196, and 217).
195. The set of aims of the members of the research community with regard to their study of the scientific domain includes searching and discovering or using regularities (in particular laws, i.e., universal patterns) and circumstances of its domain, systematising (into theories) general hypotheses which generalise information about its domain, applying those theories, and refining methods in its methodics (Tuomela, 1985, pp. 216–8; Bunge, 2003, p. 260).
196. Looks for universal knowledge/truths (cf. Bunge, 2013, pp. 21–2).
197. Looks for the composition, structure, and the kind and intensity of effects of things (cf. ibid., p. 22).
198. Epistemic humility: the psychological or at least sociological awareness or at least recognition of the fact that one might be incorrect or mistaken (Mahner, 2013, p. 39; Lilienfeld, 2016, p. 268). This is because of property 63 (the general fallibility/provisionality/tentativity of empirical science).
199. Open-mindedness (Gruenberger, 1964, p. 1415). "The self-correction of science can be short-circuited or delayed by close-mindedness" (Lilienfeld, 2016, p. 271).
200. "Genuine science embraces skepticism; pseudoscience tends to view skepticism as a sign of narrow-mindedness." (Carey, 2011, p. 127).

[7] External consistency is an unnecessary property because it could be the case that the bulk of external "knowledge" were false.

Social

201. Its research community is a social system composed of persons who have received a specialised training, hold strong communication links among themselves (Mahner, 2013, p. 38), and initiate or continue a tradition of inquiry (Bunge, 2003, p. 260). The members of that community are rational and well-enough trained to be able to perform relevant practical inferences and to act accordingly (Tuomela, 1985, pp. 216–8).
202. Its specialised community possesses the (sets of) norm(s) of universalism (Merton, 1973 [1942], pp. 270–3), which asserts that whatever their origins, truth claims should be subjected to preestablished, impersonal criteria. This implies that the acceptance or rejection of claims should not depend on the personal or social qualities of their protagonists (Hansson, 2017).
203. Its specialised community possesses the (sets of) norm(s) of "communism" (Merton, 1973 [1942], pp. 273–5), which "says that the substantive findings of science are the products of social collaboration and therefore belong to the community, rather than being owned by individuals or groups" (Hansson, 2017).
204. Its specialised community possesses the (sets of) norm(s) of disinterestedness (Merton, 1973 [1942], pp. 275–7), which "imposes a pattern of institutional control that is intended to curb the effects of personal or ideological motives that individual scientists may have" (Hansson, 2017).
205. Its specialised community possesses the (sets of) norm(s) of organised scepticism (Merton, 1973 [1942], pp. 277–8), which "implies that science allows detached scrutiny of beliefs that are dearly held by other institutions. This is what sometimes brings science into conflicts with religions and [other] ideologies." (Hansson, 2017).
206. Willingness to test: The community that proposes the theory or that practices the discipline features a high degree of disposition, willingness, and concern towards testing, confirming or disconfirming their theory (Popper, 1959/2002, p. 280; Mahner, 2013, p. 32; Lilienfeld, 2016, p. 272). Scientists are *willing to test*: if it is possible to test a theory, it will be tested (Hansson, 1983, cited in Hansson, 2017). That community does not avoid "putting its claims to a meaningful test" (Coker, 2001, p. 2), e.g., careful, methodical experiments.
207. Disposition to comply: Features a high degree of disposition of that community to comply with the results of the tests once these have been carried out (Boudry, 2013b, p. 87). The community does not "ignore results of those carried out" (Coker, 2001, p. 2).
208. Its research community does not evade scientific peer review (Lilienfeld et al., 2012, pp. 21, 24; Lilienfeld et al., 2015, pp. 7–8).
209. Features a commitment to rooting out errors in its web of beliefs (Lilienfeld et al., 2012, p. 12).
210. Features a high degree of concern in the community proposing the theory for developing solutions or legitimate explanations for its unsolved anomalies, in case these exist. The community of theorists attempts to develop the theory towards solutions to these problems (Thagard, 1978, pp. 227–8; Fernandez-

Beanato, 2017, p. 33). The possession of in-principle testability (property 51) is necessary for the possession of this property.

2.2.2.5 Specific Epistemic Properties of Practices, Propositions, Hypotheses, and Theories, and of the Evolution of the Discipline

211. All its procedurally valid studies give consistent results (Coker, 2001, p. 5).
212. Theories are evaluated historically and contextually, in relation to available alternative theories (Thagard, 1988, pp. 161–71).
213. Scientific theories are more progressive than alternative theories over a long period of time (ibid., p. 168). They perform their function better than any other theories (Dutch, 1982, pp. 9–10).
214. Features cognitive, methodological, and institutional mechanisms for the elimination of errors (cf. Mahner, 2013, p. 39).
215. Features theoretical (or programmatic) mechanisms for the elimination of anomalies, such as the mechanisms characteristic of the positive heuristic of the Lakatosian scientific research programmes. Scientific programmes tend to feature such mechanisms (ibid.).
216. The problems presented by the theory, if there are any, are solved (based on the theory's power to solve anomalies) (ibid., p. 32; Nickles, 2013, p. 116).
217. General self-correctiveness (Casti, 1990, p. 59; Lilienfeld et al., 2012, pp. 21, 23, 29; Lilienfeld et al., 2015, p. 7; cf. Lack & Rousseau, 2016, pp. 41, 47; Lilienfeld, 2016, pp. 271–2). "Science is plastic and progressive in the sense that it corrects its results (its 'truths') – and, what is more, also its methods. What is most important in science is accordingly just its self-corrective and progressive method rather than its substance or content at any given time." Self-correctiveness is "a central feature of the scientific method not possessed by any other method for gathering knowledge about the world. Critical scientific discussion, which also concerns the results of testing scientific hypotheses, can be expected to lead to a process in which inquiry corrects its own mistakes – be these mistakes in erroneous data or false theories, or even errors in the scientific method itself". "[S]cience can systematically eliminate errors. Thus it can be shown for some central scientific tests at least that, when the observations accumulate, an erroneous hypothesis is very likely to be rejected. The same can be said of scientific discussion and debate at large". "[S]elf-correction can apply both to truth and the methods of obtaining truths" (Tuomela, 1985, pp. 210–1, 213–4). The community of theorists is open to correction (Lack & Rousseau, 2016, p. 20). "[E]rroneous claims tend to be gradually ferreted out [of science] by a process akin to natural selection" (Lilienfeld et al., 2015, p. 11).
218. Is "a self-sufficient institution which does not allow external checks on validity; there are no extra-scientific criteria of [scientific] correctness" (Tuomela, 1985, p. 214).

219. Autonomy (Merton, 1973 [1942], pp. 257–9). The society or societies in which science is practiced offer/s the research community autonomy of inquiry (Tuomela, 1985, pp. 216–8; Mahner, 2013, p. 38). "[S]cience is self-corrective partly because it is autonomous [...] in the sense that science is or should be autonomous in the choice of the criteria of truth and correctness, as well as in the application of these criteria in scientific inquiry. On the other hand, science can recover from at least minor violations of its autonomy – as we can judge from history (cf., e.g., the Lysenko affair)" (Tuomela, 1985, p. 214).
220. Is not based on some alleged paranormal epistemic super-abilities of an elite (cf. ibid., p. 228).

2.3 Scoring Method

In this work each of the fields studied is assessed with respect to its scientificity or non-scientificity on the basis of the following scoring method:

- For each property and each assessed field, this work either determines that the field (mostly) possesses the property, or that it (mostly) lacks it, or that neither of these statuses can be determined from the sources used, by our best analysis.
- Regarding non-specific properties, each assessed field receives −1 "point" for each property that the field is determined to lack, 0 points for each property that it is determined to neither lack nor possess (or if it cannot be determined either way), and 0 points for each property that the field is determined to possess. This last provision is due to the fact that the possession of non-specific properties, although important, is not demarcatory, as explained above.
- Regarding specific properties, each assessed field obtains −1 point for each property that it is determined to lack, 0 points for each property that it is determined to neither lack nor possess (or if it cannot be determined either way), and +1 point for each property that it is determined to possess. This is due to the fact that, as explained above, both the lack and the possession of specific properties are demarcatory.

The positive adjudication of a property to a field will obviously not imply that every work produced in that field has the adjudicated property.

This scoring method is deliberately coarse-grained. One could assign different real numbers to different properties according to their correspondingly different degrees of importance. For example, we could have given predictive accuracy a "value" or "weight" of 1 and scientifically peer-reviewed publication only a value of 0.5. And one could assign different real numbers to the correspondingly different degrees to which we think a field possesses each property. For example, we could have given some portions of physics 1 for predictive accuracy and some other portions of physics only 0.5 for the same property. This might be interesting to do in particular cases, but would make the analysis proposed here impractical, and in any case this work's results show that introducing those complications is unnecessary to, at least in the cases covered here, successfully demarcate SBS and different types of sciences.

The following 17 chapters are dedicated to considering a wide range of clear cases[8] of particular sciences and of non-sciences of different subtypes (natural sciences, social sciences, pseudosciences, non-pseudoscientific non-sciences, etc.), including studies on all the clear cases of non-scientific fields mentioned in Sect. 2.1, all of which are assessed qua CFs.

The fact that my method has been to consider this range of different cases has necessarily meant that I have not gotten into as much depth about each individual case as I could have. In fact, a work of the extension of this entire work could be written on each of the 17 individual cases.

These case studies use the 220 low-level properties featured in this list as SD criteria. The hypothetical status of these properties as collectively demarcatory of science is substantiated via these case studies, all of which confirm the hypothesis, turning it into a promising and strong thesis.

References

Baran, G. R., Kiana, M. F., & Samuel, S. P. (2014). Chapter 2: Science, Pseudoscience, and not science: How do they differ? In G. R. Baran, M. F. Kiana, & S. P. Samuel (Eds.), *Healthcare and biomedical technology in the 21st century* (pp. 19–57). Springer.

Beyerstein, B. (1996). *Distinguishing science from pseudoscience*. The Center for Curriculum and Professional Development.

Blancke, S., Boudry, M., & Braeckman, J. (2018). Whence pseudoscience? An epidemiological approach. *Metode*. https://doi.org/10.7203/metode.8.10007

Blancke, S., Boudry, M., & Braeckman, J. (2019). Reasonable irrationality: The role of reasons in the diffusion of pseudoscience. *Journal of Cognition and Culture, 19*(5).

Borinskaya, S. A., Ermolaev, A. I., & Kolchinsky, E. I. (2019). Lysenkoism against genetics: The meeting of the Lenin all-union academy of agricultural sciences of August 1948, its background, causes, and aftermath. *Genetics, 212*(1), 1–12.

Boudry, M. (2011). Here be dragons: Exploring the hinterland of science. Ghent, Belgium: Ghent University. Faculty of Arts and Philosophy. https://biblio.ugent.be/publication/1191286

Boudry, M., & Braeckman, J. (2011). Immunizing strategies & epistemic defense mechanisms. *Philosophia, 39*(1), 145–161.

Boudry, M., & Buekens, F. (2011). The epistemic predicament of a pseudoscience: Social constructivism confronts Freudian psychoanalysis. *Theoria, 77*(2), 159–179.

Boudry, M., & Braeckman, J. (2012). How convenient! The epistemic rationale of Selfvalidating belief systems. *Philosophical Psychology, 25*(3), 341–364.

Boudry, M. (2013a). The hypothesis that saves the day. Ad Hoc reasoning in pseudoscience. *Logique et Analyse, 223*, 245–258.

[8] If the classification of any of the chosen fields appears to be less than clear to the reader (e.g., if a reader believes that parapsychology is currently not clearly pseudoscientific), then the overall results confirming that classification should add support to it. This is in line with one of the intended uses of this work, which is to serve as a tool to ascertain scientificity or non-scientificity in less clear cases.

Boudry, M. (2013b). Loki's Wager and Laudan's error. On genuine and territorial demarcation. In M. Pigliucci & M. Boudry (Eds.), *Philosophy of pseudoscience: Reconsidering the demarcation problem* (pp. 79–98). The University of Chicago Press.
Boudry, M. (2019). "Diagnosing pseudoscience in real life (a tribute to Sven Ove Hansson)". [Preprint], in http://philsci-archive.pitt.edu/18630/
Boudry, M. (2022). Diagnosing pseudoscience—by getting rid of the demarcation problem. *Journal for General Philosophy of Science, 53*(2), 83–101.
Brigandt, I. (2013). Intelligent design and the nature of science: Philosophical and pedagogical points. In K. Kampourakis (Ed.), *History, philosophy and theory of the life sciences* (Vol. 1, pp. 205–238).
Bunge, M. A. (1960). *La ciencia: Su método y su filosofía*. EUDEBA.
Bunge, M. A. (1982). Demarcating science from pseudoscience. *Fundamenta Scientiae, 3*, 369–388.
Bunge, M. A. (1984). What is pseudoscience? *Skeptical Inquirer, 9*, 36–46.
Bunge, M. A. (2003). *Philosophical dictionary* (Enlarged ed.). Prometheus Books.
Bunge, M. A. (2013). *Medical philosophy: Conceptual issues in medicine*. World Scientific.
Carey, S. S. (2011). *A beginner's guide to scientific method* (4th ed.) Wandsworth.
Casti, J. L. (1990). *Paradigms lost*. Harper Perennial.
Cleland, C. E., & Brindell, S. (2013). Science and the Messy, uncontrollable world of nature. In M. Pigliucci & M. Boudry (Eds.), *Philosophy of pseudoscience: Reconsidering the demarcation problem* (pp. 183–202). The University of Chicago Press.
Coker, R. (2001). Distinguishing science and pseudoscience. http://hep.physics.utoronto.ca/~orr/wwwroot/JPH441/Pseudoscience.pdf
Derksen, A. A. (1993). The seven sins of pseudo-science. *Journal for General Philosophy of Science, 24*, 17–42.
Derksen, A. A. (2001). The seven strategies of the sophisticated pseudo-scientist: A look into Freud's rhetorical tool box. *Journal for General Philosophy of Science, 32*, 329–350.
Dutch, S. I. (1982). Notes on the nature of fringe science. *Journal of Geological Education, 30*, 6–13.
Erduran, S., & Dagher, Z. R. (2014). *Reconceptualizing the nature of science for science education: Scientific knowledge, practices and other family categories*. Springer.
Fales, E. (2013). Is a science of the supernatural possible? In M. Pigliucci & M. Boudry (Eds.), *Philosophy of pseudoscience: Reconsidering the demarcation problem* (pp. 247–262). The University of Chicago Press.
Fasce, A. (2017). What do we mean when we speak of pseudoscience? The development of a demarcation criterion based on the analysis of twenty-one previous attempts. *Disputatio. Philosophical Research Bulletin, 6*(7), 459–488.
Faulk, C. (2013). Lamarck, Lysenko, and modern day epigenetics. Mind the Science Gap. http://www.mindthesciencegap.org/2013/06/21/lamarck-lysenko-and-modern-day-epigenetics/
Fernandez-Beanato, D. A. (2017). Non ergo arte: *Demarcacionismo entre ciencia y no-ciencia en la Antigüedad a través de obras de Cicerón y Agustín de Hipona*. Unpublished Master's thesis. Universidad Nacional de Tres de Febrero, Tres de Febrero, Argentina.
Fernandez-Beanato, D. (2021). Feng Shui and the demarcation project. *Science & Education, 30*(6), 1333–1351.
Glymour, C., & Stalker, D. (1990). Winning through pseudoscience. In P. Grim (Ed.), *Philosophy of science and the occult* (2nd ed., pp. 92–103). State University of New York Press.
Goode, E. (2013). Paranormalism and pseudoscience as deviance. In M. Pigliucci & M. Boudry (Eds.), *Philosophy of pseudoscience: Reconsidering the demarcation problem* (pp. 145–164). The University of Chicago Press.
Grim, P. (1990a). Astrology. introduction. In P. Grim (Ed.), *Philosophy of science and the occult* (2nd ed., pp. 15–17). State University of New York Press.
Grim, P. (1990b). Parapsychology. Introduction. In P. Grim (Ed.), *Philosophy of science and the occult* (2nd ed., pp. 183–186). State University of New York Press.

Grove, J. W. (1985). Rationality at risk: Science against pseudoscience. *Minerva, 23*, 216–240.
Gruenberger, F. J. (1964). A measure for crackpots. *Science, 145*, 1413–1415.
Hansson, S. O. (1983). *Vetenskap och ovetenskap*. Tiden.
Hansson, S. O. (2013). Defining pseudoscience and science. In M. Pigliucci & M. Boudry (Eds.), *Philosophy of pseudoscience: Reconsidering the demarcation problem* (pp. 61–78). The University of Chicago Press.
Hansson, S. O. (2017). Science and pseudo-science. In E. N. Zalta (Ed.), *The Stanford Encyclopedia of Philosophy* (Summer 2017 ed.). Stanford University. https://plato.stanford.edu/archives/sum2017/entries/pseudo-science/
Harker, D. (2017). Demarcation and the created controversy. *Philosophia, 45*(1), 1–10.
Hempel, C. G. (1942). The function of general laws in history. *Journal of Philosophy, 39*, 35–48. Reprinted in Hempel, C. G., 1965, pp. 231–244.
Hempel, C. G. (1951). The concept of cognitive significance: A reconsideration. *Proceedings of the American Academy of Arts and Sciences, 80*, 61–77.
Hempel, C. G. (1965). *Aspects of scientific explanation and other essays in the philosophy of science*. Free Press.
Hempel, C. G., & Oppenheim, P. (1948). Studies in the logic of explanation. *Philosophy of Science, 15*, 135–175. Reprinted in Hempel, C. G., 1965, pp. 245–290.
Hirvonen, I., & Karisto, J. (2022). Demarcation without dogmas. *Theoria, 88*(3), 701–720.
Hoyningen-Huene, P. (2013). *Systematicity: The nature of science*. Oxford University Press.
Irzik, G., & Nola, R. (2014). New directions for nature of science research. In M. R. Matthews (Ed.), *International handbook of research in history, philosophy and science teaching* (pp. 999–1021). Springer.
Kahneman, D. (2011). *Thinking, fast and slow*. Farrar, Straus and Giroux.
Kitcher, P. (1982). *Abusing science. The case against creationism*. MIT Press.
Koertge, N. (2013). Belief buddies versus critical communities: The social organization of pseudoscience. In M. Pigliucci & M. Boudry (Eds.), *Philosophy of pseudoscience: Reconsidering the demarcation problem* (pp. 165–180). The University of Chicago Press.
Kuhn, T. S. (1977). Objectivity, value judgment, and theory choice. In T. Kuhn (Ed.), *The essential tension: Selected studies in scientific tradition and change* (pp. 320–339). The University of Chicago Press.
Kuhn, T. S. (1996). *The structure of scientific revolutions* (3rd ed.). University of Chicago Press.
Lack, C., & Rousseau, J. (2016). *Critical thinking, science, and pseudoscience: Why we can't trust our brains*. Springer.
Ladyman, J., Ross, D., Spurrett, D., & Collier, J. G. (2007). *Everything must go: Metaphysics naturalised*. Oxford University Press.
Ladyman, J. (2013). Toward a demarcation of science from pseudoscience. In M. Pigliucci & M. Boudry (Eds.), *Philosophy of pseudoscience: Reconsidering the demarcation problem* (pp. 45–60). The University of Chicago Press.
Lakatos, I. (1970). Falsification and the methodology of scientific research programmes. In I. Lakatos & A. Musgrave (Eds.), *Criticism and the growth of knowledge* (pp. 91–195). Cambridge University Press.
Lakatos, I. (1978). *The methodology of scientific research programmes*. Cambridge University Press.
Langmuir, I. ([1953] 1989), "Pathological science", *Physics Today, 42*(10), 36–48.
Laudan, L. (1983). The demise of the demarcation problem. In R. S. Cohen & L. Laudan (Eds.), *Physics, philosophy and psychoanalysis* (pp. 111–127). Springer Netherlands.
Law, S. (2011). *Believing Bullshit: How not to get sucked into an intellectual black hole*. Prometheus Books.
Law, S. (2019). How can we tell science from pseudoscience? In K. McCain & K. Kampourakis (Eds.), *What is scientific knowledge?* (pp. 100–116). Routledge.
Lawson, T. J. (2007). *Scientific perspectives on pseudoscience and the paranormal: Readings for general psychology*. Pearson Prentice Hall.

Lilienfeld, S., Ammirati, R., & David, M. (2012). Distinguishing science from pseudoscience in school psychology: Science and scientific thinking as safeguards against human error. *Journal of School Psychology, 50*(1), 7–36.
Lilienfeld, S. O., Lynn, S. J., & Lohr, J. M. (2015). *Science and pseudoscience in clinical psychology* (2nd ed.). The Guilford Press.
Lilienfeld, S. (2016). Afterword: Science and humility. In C. Lack & J. Rousseau (Eds.), *Critical thinking, science, and pseudoscience: Why we can't trust our brains* (pp. 267–276). Springer.
Mahner, M. (2007). Demarcating science from non-science. In T. Kuipers (Ed.), *Handbook of the philosophy of science: General philosophy of science – Focal issues* (pp. 515–575). Elsevier.
Mahner, M. (2013). Science and pseudoscience. How to demarcate after the (alleged) demise of the demarcation problem. In M. Pigliucci & M. Boudry (Eds.), *Philosophy of pseudoscience: Reconsidering the demarcation problem* (pp. 29–44). The University of Chicago Press.
Mc Mullin, E. (1993). Evolution and special creation. *Zygon, 28*(3), 299–335.
Merton, R. K. [1942] (1973). Science and technology in a democratic order, *Journal of Legal and Political Sociology, 1*, 115–126, 1942. Reprinted as "The Normative Structure of Science". In R. K. Merton, *The sociology of science. Theoretical and empirical investigations* (pp. 267–278). Chicago: University of Chicago Press.
Mukerji, N., & Ernst, E. (2022). Why homoeopathy is pseudoscience. *Synthese, 200*, 394.
Nagel, E. (1961). *The structure of science: Problems in the logic of scientific explanation*. Harcourt, Brace & World.
Nickles, T. (2013). The problem of demarcation. History and future. In M. Pigliucci & M. Boudry (Eds.), *Philosophy of pseudoscience: Reconsidering the demarcation problem* (pp. 101–120). The University of Chicago Press.
Oreskes, N. (2004). The scientific consensus on climate change. *Science, 306*(5702), 1686.
Overton, W. (1982). Creationism in schools: The decision in McLean vs. the Arkansas Board of Education. *Science, 215*(4535), 934–943.
Park, R. (2003), *Seven warning signs of bogus science*. http://www.unl.edu/rhames/park-seven-signs.pdf
Pigliucci, M., & Boudry, M. (2013). Introduction. In M. Pigliucci & M. Boudry (Eds.), *Philosophy of pseudoscience: Reconsidering the demarcation problem* (pp. 1–6). The University of Chicago Press.
Pigliucci, M. (2013). The demarcation problem. A (belated) response to Laudan. In M. Pigliucci & M. Boudry (Eds.), *Philosophy of pseudoscience: Reconsidering the demarcation problem* (pp. 9–28). The University of Chicago Press.
Pigliucci, M., & Boudry, M. (2014). Prove it! The burden of proof game in science vs pseudoscience disputes. *Philosophia, 42*(2), 487–502.
Popper, K. (1959/2002). *The logic of scientific discovery*, translation of *Logik der Forschung*. Repr. London: Hutchinson.
Popper, K. (1962). *Conjectures and refutations: The growth of scientific knowledge*. Basic Books.
Radner, D., & Radner, M. (1982). *Science and unreason*. Wadsworth.
Ruse, M. (Ed.). (1996). *But is it science? The philosophical question in the creation/evolution controversy*. Prometheus Books.
Ruse, M. (2013). Evolution. From pseudoscience to popular science, from popular science to professional science. In M. Pigliucci & M. Boudry (Eds.), *Philosophy of pseudoscience: Reconsidering the demarcation problem* (pp. 225–246). The University of Chicago Press.
Russell, B. (1949). *The scientific outlook* (2nd ed.). George Allen & Unwin Ltd.
Sagan, C. (1976). "Objections to astrology" (letter to the editor). *The Humanist, 36*, 2.
Schick, T., & Vaughn, L. (1995). *How to think about weird things. Critical thinking for a new age*. McGraw-Hill.
Schindler, S. (2018). *Theoretical virtues in science: Uncovering reality through theory*. Cambridge: Cambridge University Press.
Shelton, J. W. (2004). *Homeopathy: How it really works*. Prometheus Books.

Shermer, M. (2013). Science and pseudoscience. The difference in practice and the difference it makes. In M. Pigliucci & M. Boudry (Eds.), *Philosophy of pseudoscience: Reconsidering the demarcation problem* (pp. 203–224). The University of Chicago Press.

Simon, M. A. (1969). When is a resemblance a family resemblance? *Mind, 78*(311), 408–416.

Skelton, R. (2011). *A survey to the forensic sciences*. Lulu.

Smith, K. (2012). Homeopathy is unscientific and unethical. *Bioethics, 26*(9), 508–512.

Sommer, A. (2014). Psychical research in the history and philosophy of science. An introduction and review. *Studies in History and Philosophy of Biological and Biomedical Sciences, 48*, 38–45.

Thagard, P. (1978). Why astrology is a pseudoscience. *PSA, 1*, 223–234.

Thagard, P. (1988). *Computational philosophy of science*. MIT Press.

Thagard, P. (2011). Evolution, creation, and the philosophy of science. In R. S. Taylor & M. Ferrari (Eds.), *Epistemology and science education: Understanding the evolution vs. intelligent design controversy* (pp. 20–37). Routledge.

Tuomela, R. (1985). *Science, action, and reality*. Reidel.

Tuomela, R. (1987). Chapter 4: Science, protoscience, and pseudoscience. In J. C. Pitt & M. Pera (Eds.), *Rational changes in science: Essays on scientific reasoning. Boston studies in the philosophy of science* (Vol. 98) (pp. 83–101). Springer.

Vollmer, G. (1993). *Wissenschaftstheorie im Einsatz, Beiträge zu einer selbstkritischen Wissenschaftsphilosophie*. Hirzel Verlag.

Part II
Case Studies

Chapter 3
The Properties of Genethliac Astrology

Abstract This chapter provides a brief definition and a description of/introduction to genethliac astrology. It then meta-analytically shows that the demarcation criteria and the scoring method proposed in Chap. 2 demarcate genethliac astrology as non-scientific, in agreement with the current philosophical and scientific consensus.

Keywords Astrology · Divination · Horoscope · Scientific demarcation · Non-science · Pseudoscience

3.1 Introduction

Let us now see how the list of properties of scientificity compiled in Sect. 2.2 can be used to rationally reconstruct, or explicate, the demarcatory consensus regarding the scientificity or non-scientificity of a wide variety of fields.

Let us begin with genethliac astrology (GA), a CF that in the academic literature on SD has been used profitably as a clear (and paradigmatic, in the everyday, not the Kuhnian sense of this word) case of a CF about which there exists a consensus among philosophers, scientists, and historians of science to identify it as non-scientific and as pseudoscientific (Thagard, 1978, p. 223; Grim, 1990, p. 15; Pigliucci, 2013, p. 17).

This chapter shows that my demarcatory procedure demarcates GA as non-scientific, in agreement with the current philosophical and scientific consensus. This section defines GA. Section 3.2 lists properties of scientificity that GA has. Section 3.3 lists properties of scientificity that GA lacks. As the conclusion of this chapter, Sect. 3.4 argues that the demarcation criteria listed in Sect. 2.2 demarcate GA as non-scientific (that GA does not have enough of the properties of scientificity, and therefore should be deemed non-scientific), and that the fact that GA is commonly deemed non-scientific is explained by the fact that it does not have enough of the properties of science listed in Chap. 2.

I take the properties of science listed in Sect. 2.2, and show that GA lacks many of them. This provides: a) a rational reconstruction (and therefore an explanation) of

D. Fernandez-Beanato, *A Working Scientific Demarcation*, Synthese Library 519, https://doi.org/10.1007/978-3-031-99005-2_3

why GA is non-scientific, b) further evidence of the extensional adequacy of this work, c) further evidence of this work's explanatory power, and d) further evidence of the fact that the proposed set of criteria demarcates science appropriately.

Our analysis of GA must accept and deal with the fact that there is a plurality of genethliac astrological theories, not all of which seem to be consistent with each other (Bennett et al., 2013, p. 82). And there are no professional institutions that are widely recognised by the international scientific or philosophical communities as "official" or orthodox representatives of GA. Therefore, this work cannot present a theory of astrology as *the* theory of GA, but rather only a common-denominator basic set of hypotheses.

Astrology is not merely the general idea that what happens in the extraterrestrial portion of the universe can affect life on Earth in non-obvious ways (less obvious than, for example, an asteroid impact). This general idea is what Feyerabend was talking about when he defended "astrology" (1978), and is true (e.g., the effects of the Moon and the Sun on Earth's tides) (Eisler, 1946, pp. 138–45, 175–6; Jerome, 1977, pp. 106, 151–63). "Science has […] come to some understanding of a number of external agents that affect this planet and its creatures in a variety of ways. […] these phenomena are very definitely non astrological" (Culver & Ianna, 1988, pp. 181–200, 216).

GA is a discipline or family of disciplines used to indicate and predict the different characteristics or traits of people, and the fortune or different types of eventualities or events that are likely to occur to people, based on the temporal (and/or geographical) circumstances of their respective births. In order to do this, it uses allegedly discovered correlations between the movements and arrangements of celestial bodies relative to each other as perceived from the location of a given birth on Earth, on the one hand, and the possession by the individual born of different traits or characteristics and probable futures, on the other (Bok & Jerome, 1975, p. 24; Brandt Riske, 2010, p. 30).

The celestial bodies that astrology takes into account are stars, the eight planets in the Solar System minus the Earth, dwarf planets in the Solar System (especially Pluto and Ceres, which can be considered astrological "planets"), the Sun, the Moon, and sometimes small Solar System bodies (e.g., Chiron).

According to GA, for any subset (that is contemplated in the astrological theory) of a set of characteristics/traits and types of probable eventualities (which are also contemplated in the astrological theory), there is a particular combination of celestial elements that determines a complete horoscope—also known as a natal chart—. This horoscope "reflects" those characteristics and the greater probability of experiencing those types of eventualities for individuals born at the time and place corresponding to said celestial combination (Brandt Riske, 2010, pp. 29, 36, 38–9). Thus, these "reflections" can be "read", "seeing" in the sky or in the natal chart earthly realities and probabilities.

To avoid committing the strawman fallacy, in the investigation whose results are presented here, I was concerned only with some of the more sophisticated, allegedly respectable pro-astrology authors, some of whom are or were recognised leaders of their field, and whose works are serious enough to have been addressed in works published by recognised philosophers and scientists (see bibliography).

Let us see how GA stands regarding the possession of our scientificity properties.

3.2 The Properties of GA

We can recognise the following properties in some versions of GA: 16, 22–23, 25–29, 31, 34–40, 42–43, 46–49, 53, 55, 62, 66–70, 76, 87, 89, 94, 106–107, 110, 116, 169, 178, 181–182, 193, 195 (except for refining its methods), 202–203, and 220.

Because I am "conceding" here (i.e., recognising and noting facts that go "against" the strength of my hypothesis), it is not necessary for our purposes to justify this recognition. The claims that substantiate my hypothesis are the ones that are justified (in the next section).

3.3 The Properties that GA (Mostly) Does Not Have

GA does not take "responsibility to establish the truth of its claims" (cf. Kelly et al., 1990, p. 73) (lack of properties 3, 39, 69, 194, and 196). It often does not check evidence (ibid., pp. 70–2). The problems presented by the theory are not solved (lack of property 216). The theories of GA have many unsolved problems (lack of property 191), yet its community of practitioners and theoreticians makes little attempt to develop their theories towards solutions to these problems (lack of property 210), shows no concern for attempts to evaluate their theories in relation to non-astrological theories (lack of property 212), and is selective in considering confirmations and disconfirmations (lack of property 78), while still calling their discipline "scientific". It does not feature a high degree of disposition, willingness, and concern in the community that proposes the theory or that practices the discipline towards testing, confirming or disconfirming their theory (lack of property 206). The community proposing the theory does not change its mind when the evidence compels it to do so (lack of properties 2, 75, 199, and 217). It does not abide by evidence (lack of property 96). There is not much care within the discipline for the question as to whether it works (Jerome, 1977, pp. 137–146; Thagard, 1978, pp. 227–8; Culver & Ianna, 1988, pp. 64, 80, 203) (lack of properties 19 and 20). There is no non-arbitrary justification for explaining similarities by using GA and dissimilarities by using hereditary or environmental factors (James, 1990, pp. 32–3; e.g., Weingarten, 1974, p. 66) (lack of property 192).

Among the aforementioned problems are:

(i) There is insufficient evidence for it (Jerome, 1977, pp. 137, 141; Kelly, 1997, p. 1062).

(ii) Its predictions are unsuccessful (Thagard, 1978, p. 227; Kurtz & Fraknoi, 1985; Culver & Ianna, 1988, pp. 10, 158, cf. p. 168; Dean & Kelly, 2003). For example, twins are different both in character/personality and in their life stories (Kelly et al., 1990, p. 52). This problem has been extant for thousands of years (cf. e.g., Cicero, 2011 [44 BCE], II.87–99; Augustine of Hippo, 1996 [c. 397–400], VII.6; 1924 [426], V.1–6).
(iii) There is no consensus within the discipline on important points.

GA has not changed much (lack of properties 14 and 176). It has been less progressive than alternative theories over a long period of time (lack of properties 84 and 213). It "is dramatically unprogressive, in that it has changed little and has added nothing to its explanatory power since the time of Ptolemy" (AD c. 100 – c. 170). It does not feature empirical progress (other than accommodation to discoveries) even when putative solutions to its unsolved anomalies or problems are offered (Eisler, 1946, p. 242; Bok & Jerome, 1975, p. 28; quote: Thagard, 1978, p. 228; Culver & Ianna, 1988, pp. 158–60, 171, 203) (lack of property 217). It does not explain all the relevant data and has not explained new classes of facts (lack of property 42). Old concepts have not been modified or discarded in light of new discoveries (Culver & Ianna, 1988, pp. 10, 150, 203) (lack of property 84).

GA does not feature a high degree of specificity in the theoretical description of the causal mechanisms that its theories propose. There is no consensus about mechanism (Eisler, 1946, pp. 186–8; Bok & Jerome, 1975, p. 31; Jerome, 1977, p. 147; Culver & Ianna, 1988, p. 37) (lack of property 179).

James (1990, pp. 28–9) claimed that because GA does work on trying to solve some of its puzzles (e.g., by including constellations in addition to the traditional 12 (e.g., Rachleff, 1973)), Kuhn's SD criteria fail to select GA as non-scientific. But the latter Kuhn (1993) admitted the existence of non-puzzle-solving sciences. Even if that were not the case, James failed to understand Kuhnian puzzle-solving, which features known theoretical accuracy of fit to nature—which is what today we call empirical adequacy—, external consistency, and fruitfulness of research findings of "new phenomena or previously unnoted relationships among those already known" (Kuhn, 1977, pp. 321–23); in other words, our properties 82, 183, and 190, all of which are conspicuously absent in GA. GA lacks empirical adequacy when some of its descriptions or predictions that are first rough approximations are statistically tested (Culver & Ianna, 1988, p. 10). GA lacks property 51 with respect to its testability in its complete form as applied to humans, as a "complete test of the astrological gestalt would involve at least 10^{35} possible combinations of astrological points, planets, signs, houses, and aspects" (ibid., pp. 150–1),[1] the human population being only 8 billion, leaving most of the astrological correspondences untested and therefore making impossible to test GA as a whole (cf. Jerome, 1977, p. 73).

Claimed astrological phenomena are at the very limit of detection, or beyond it, as in the Gauquelin case (cf. Culver & Ianna, 1988, p. 141) (lack of property 61).

[1] This assessment was made before the discoveries of dwarf planets Eris, Makemake, and Haumea, and of many other bodies in the Solar System.

GA is vague. "Astrologists never describe precisely what their model predicts, what the criteria are by which it could be tested, and what sort of evidence they would accept as proving it wrong" (Culver & Ianna, 1988, pp. 10, 174, 220; quote: Kelly et al., 1990, p. 62) (lack of property 92). Being vague can even be presented as an ethical imperative within GA (e.g., Davison, 1963, p. 226).

GA ignores or denies negative or conflicting evidence or failures (Culver & Ianna, 1988, p. 174; Kelly et al., 1990, p. 71) (lack of properties 13–14, 75, 96, and 120) and does not use the general scientific method (Culver & Ianna, 1988, pp. 201, 203) (lack of properties 103–104).

It features appeals to personal authority (e.g., Davison, 1963, p. 15; West & Toonder, 1973 [1970], pp. 81–92; James, 1990, p. 32) (lack of property 108).

It uses ad hoc hypotheses to immunise claims from falsification (Culver & Ianna, 1988, p. 99), including built-in subterfuge (Kelly et al., 1990, p. 73) (lack of properties 10, 12, and 185). For example, "if a person whose horoscope predicts [good] fortune instead encounters failure, the astrologer seeks to explain this by subsuming the relevant horoscope under a more inclusive one that calls for misfortune" (James, 1990, p. 28). For example, subsuming the horoscope of a person under that of her city of residence, or under that of her country. This is done by virtue of the "theory of subsumption" (West & Toonder, 1973 [1970], pp. 128, 148; Weingarten, 1974, p. 65), which indicates "that horoscopes with a narrower remit are contained within ones with a wider relevance. The horoscope of an individual is contained within that for the family, the wider community [...]" (Burns, 2018, p. 242).

It suffers from self-confirmation bias (Culver & Ianna, 1988, pp. 150–1; Kelly et al., 1990, p. 60) and other cognitive errors and biases. "[A]strologers are generally unaware of the many hidden persuaders that can make them see hits where none exist" (Dean et al., 1996, pp. 89–93) (quote: Dean & Kelly, 2003, pp. 180–1) (lack of property 147).

It uses fallacies (James, 1990, p. 32) (lack of property 129).

It excessively relies on anecdotes, and sometimes considers that anecdotes trump (other kinds of) evidence (James, 1990, p. 32; e.g., Davison, 1963, pp. 14–5; West & Toonder, 1973 [1970], pp. 282–4) (lack of properties 132–133 and 145).

It advocates or markets an unproven practice (Culver & Ianna, 1988, p. 10) (lack of property 135).

It does "research by exegesis" (e.g., West & Toonder, 1973 [1970], pp. 1–4) (lack of property 137).

It does not resist the allure of seeing sense in spectacular coincidences (James, 1990, p. 32; e.g., Davison, 1963, pp. 14–5; West & Toonder, 1973 [1970], pp. 282–4) (lack of property 143).

GA often uses intuition as a means for subjective validation (Kelly et al., 1990, pp. 70–2) (lack of property 156).

It uses resemblance thinking (Thagard, 1988, pp. 166–8) and appeals to magical thinking (Eisler, 1946, pp. 138, 161–7, 169, 174, 178–86, 244; Jerome, 1977, pp. 2, 20–32, 69–78, 95–7; e.g., Davison, 1963, p. 34: "The name of each planet [even of the ones that were named in the Modern era] gives a clue to its [astrological] nature") (lack of property 171).

It appeals to myths (Culver & Ianna, 1988, p. 94; e.g., Davison, 1963, pp. 34, 41–3) (lack of property 173).

It is irrational (James, 1990, pp. 32–5) (lack of property 65).

GA is not scientifically peer-reviewed in scientific publications (astrologists do perform peer-review of their peers; and GA can be, and is, examined by science). GA is not subject to scientific peer review (lack of property 128) and avoids scientific peer review (lack of property 208) by having its own publications.

The sources used for this assessment also make clear that GA's specialised community lacks the (sets of) norm(s) of organised scepticism (lack of property 205); that it does not feature cognitive, methodological, and institutional mechanisms for the elimination of errors (lack of property 214); and that it does not feature theoretical (or programmatic) mechanisms for the elimination of anomalies (lack of property 215).

Regarding all the properties in the demarcatory list not mentioned in this chapter, it is unclear whether GA mostly has them.

3.4 Chapter Conclusions

GA lacks at least 56 of the properties of scientificity featured in the demarcatory list, while featuring at least 16 of the specific ones. This yields a score of −40 for GA. This (possibly provisional) tally is strongly against the scientificity of GA, and is the best explanation for the fact that GA is commonly deemed non-scientific.

If by using the method for SD proposed here we can always demarcate science as intended (i.e., according to the currently existing international philosophical and scientific consensus regarding the extension of the term "science"), then this method is bound to select GA as a non-scientific field (because that is the consensus).

The account offered here is empirical. Chapters 3–19 show that, out of the consensually clear cases of scientific CFs, even those with the lowest scientificity scores score far higher than the highest-scoring clear cases of non-scientific CFs. As indicated in Chap. 1, those clear cases are the ones that will serve us to establish the cut value for the demarcation of science.

By repeating this exercise in the following chapters with a wide range of consensually clear cases of particular sciences and of non-sciences of different subtypes (natural sciences, social sciences, pseudosciences, non-pseudoscientific non-sciences, etc.), I aim to find out a rank ordering of values that is consistent with the above-mentioned consensus.

References

Augustine of Hippo (Aurelius Augustinus). (1996 [c. 397–400]). *Confessionum*. Stutgardiae: Teubner.

Augustine of Hippo (Aurelius Augustinus). (1924 [426]). *De civitate Dei contra paganos : libri XXII*. Lipsiae: Teubner.

Bennett, J., Donahue, M., Schneider, N., & Voit, M. (2013). *The cosmic perspective* (7th ed.). Pearson Education Limited.

Bok, B. J., & Jerome, L. E. (1975). *Objections to astrology*. Prometheus Books.

Brandt Riske, K. (2010). *Llewellyn's complete book of astrology: The easy way to learn astrology* (1st electronic ed.). Llewellyn Publications.

Burns, W. E. (Ed.). (2018). *Astrology through history: Interpreting the stars from ancient Mesopotamia to the present*. ABC-CLIO.

Cicero, M. T. (2011 [44 BCE]). M. Tulli Ciceronis scripta quae manserunt omnia (ed. stereot. ed. primae (1938). Reprint 2011 ed. De divinatione. De fato. Timaeus. Bibliotheca scriptorum Graecorum et Romanorum Teubneriana, Vol. fasc. 46) (W. Ax & O. Plasberg, eds.). : De Gruyter.

Culver, R. B., & Ianna, P. A. (1988). *Astrology: True or false?* Prometheus Books.

Davison, R. C. (1963). *Astrology*. Bell.

Dean, G., Mather, A., & Kelly, I. W. (1996). Astrology. In G. Stein (Ed.), *The encyclopedia of the paranormal* (pp. 47–99). Prometheus Books.

Dean, G., & Kelly, I. W. (2003). Is astrology relevant to consciousness and psi? *Journal of Consciousness Studies, 10*, 175–198.

Eisler, R. (1946). *The Royal art of astrology*. H. Joseph.

Feyerabend, P. (1978). The strange case of astrology. In P. Feyerabend (Ed.), *Science in a free society* (pp. 91–96). NLB.

Grim, P. (1990). Astrology. Introduction. In P. Grim (Ed.), *Philosophy of science and the occult* (2nd ed., pp. 15–17). State University of New York Press.

James, E. W. (1990). On dismissing astrology and other irrationalities. In P. Grim (Ed.), *Philosophy of science and the occult* (2nd ed., pp. 28–36). State University of New York Press.

Jerome, L. E. (1977). *Astrology disproved*. Prometheus Books.

Kelly, I. W., Dean, G. A., & Saklofske, D. H. (1990). Astrology: A critical review. In P. Grim (Ed.), *Philosophy of science and the occult* (2nd ed., pp. 51–81). State University of New York Press.

Kelly, I. W. (1997). Modern astrology: A critique. *Psychological Reports, 81*, 1035–1066.

Kuhn, T. S. (1977). Objectivity, value judgment, and theory choice. In *The essential tension: Selected studies in scientific tradition and change* (pp. 320–339). The University of Chicago Press.

Kuhn, T. S. (1993). Afterwords. In P. Horwich (Ed.), *World changes. Thomas Kuhn and the nature of science* (pp. 311–341). MIT Press.

Kurtz, P. & Fraknoi, A. (1985). Scientific tests of astrology do not support its claims. *Skeptical Inquirer*, Spring 1985, (pp. 210–211).

Pigliucci, M. (2013). The demarcation problem. A (belated) response to Laudan. In M. Pigliucci & M. Boudry (Eds.), *Philosophy of pseudoscience: Reconsidering the demarcation problem* (pp. 9–28). The University of Chicago Press.

Rachleff, O. S. (1973). *Sky diamonds*. Hawthorne Books.

Thagard, P. (1978). Why astrology is a pseudoscience. *Proceedings of the Biennial Meeting of the Philosophy of Science Association, 1*, 223–234.

Thagard, P. (1988). *Computational philosophy of science*. MIT Press.

Weingarten, H. (1974). *A modern introduction to astrology*. ASI.

West, J. A. & Toonder, J. G. (1973 [1970]). *The case for astrology*. Baltimore, Maryland: Penguin Books Inc.

Chapter 4
The Properties of Parapsychology

Abstract This chapter provides a brief definition and a description of/introduction to parapsychology. It then meta-analytically shows that the demarcation criteria and the scoring method proposed in Chap. 2 demarcate parapsychology as non-scientific, in agreement with the current philosophical and scientific consensus.

Keywords Parapsychology · Paranormal · Scientific demarcation · Non-science · Pseudoscience · Scientificity

4.1 Introduction

This chapter shows that the demarcation criteria demarcate parapsychology as non-scientific, in agreement with the current philosophical and scientific consensus. This section defines parapsychology. Section 4.2 lists properties of scientificity that parapsychology has. Section 4.3 lists properties of scientificity that parapsychology lacks. As the conclusion of this chapter, Sect. 4.4 argues that the demarcation criteria listed in Sect. 2.2 demarcate parapsychology as non-scientific (that parapsychology does not have enough of the properties of scientificity, and therefore should be deemed non-scientific), and that the fact that parapsychology is commonly deemed non-scientific is explained by the fact that it does not have enough of the properties of science listed in Chap. 2.

I take the properties of science listed in Sect. 2.2, and show that parapsychology lacks many of them. This provides: a) a rational reconstruction (and therefore an explanation) of why parapsychology is non-scientific, b) further evidence of the extensional adequacy of this work, c) further evidence of this work's explanatory power, and d) further evidence of the fact that the proposed set of criteria demarcates science appropriately.

D. Fernandez-Beanato, *A Working Scientific Demarcation*, Synthese Library 519, https://doi.org/10.1007/978-3-031-99005-2_4

In this work, "paranormal" means "violating certain basic limiting principles for the description and explanation of reality that are currently accepted by clearly scientific fields".[1]

Parapsychology, in the narrow sense, is the study of alleged anomalies that are paranormally explainable by extrasensory[2] perception (ESP, also known as psi-gamma) or by psychokinesis (also known as telekinesis, and as psi-kappa; i.e., the alleged ability of a mind to influence an external physical system without any known physical interaction). The concept of ESP includes telepathy (direct "mind-to-mind communication"), clairvoyance, paranormal remote viewing, paranormal retrocognition, paranormal precognition, psychometry (i.e., psychically obtaining information via touching objects, also known as psychoscopy), and automatic writing.

Parapsychology receives its name because it studies alleged phenomena that are explainable by paranormal ("para-") mental activity ("psychology"). But it is not the case that, within parapsychology, these alleged anomalies are deemed paranormal by definition, because it is not necessary to believe that they are paranormal in order to consider them to be worth studying just in case they are. The studies of the alleged anomalies that are explainable by ESP or by psychokinesis are commonly grouped as "psi" studies, and ESP and psychokinesis themselves are commonly grouped as "psi" alleged types of occurrences (Reinsel, 1990, p. 201).

Parapsychology in the broad sense is a wide field that also includes the study of alleged anomalies that are explainable by/as other alleged paranormal types of occurrences.[3] For manageability reasons, this assessment, from now on, only considers the restricted meaning of the term "parapsychology" that refers exclusively to the field dealing only with psi.

Parapsychology (at least the type of parapsychology that resorts to paranormal explanations) is another CF commonly considered to be non-scientific and pseudoscientific (Grim, 1990, pp. 183, 186; Sommer, 2014, p. 42) by both philosophers and scientists.

Let us see how parapsychology stands regarding the possession of our scientificity properties.

---

[1] Other than parapsychology, a field that features paranormal explanations is ufology. Cryptozoology is sometimes counted as a discipline dealing with the "paranormal" (using varied meanings of the word "paranormal"), although an exclusively normal and broadly scientific cryptozoology, be it a subfield of zoology or an intersection between folklore studies and zoology, can easily be imagined. All kinds of divination, such as astrology (which is covered in Chap. 3), dowsing, palmistry (also known as chiromancy), and the tarot, are based on beliefs in the paranormal. Another alleged paranormal phenomenon, usually interpreted religiously, is stigmatisation.

[2] Here, "extrasensory" means "without the use of any sense recognised by currently established science", not "without the use of any sense".

[3] E.g., the electronic voice phenomenon, astral projection/travel, reincarnation, apparitional experiences, poltergeists, levitation, bilocation, mediumistic phenomena, materialization, dematerialization, auras (also known as human energy fields), xenoglossy (also known as xenoglossia, and as xenolalia), psychic healing or surgery, psychic photography (also known as thoughtography, and as projected thermography), regression to an earlier time.

4.2 The Properties of Parapsychology

We can concede that the best parapsychological approaches have the following properties: 3, 7–8, 19–20, 22–23, 29–31, 34–36, 38–40, 42–43, 47, 49, 51–54, 56, 62–63, 66–69, 71, 73, 85, 107, 110, 119, 129, 133, 138, 150, 160, 171, 194–196, 202–203, 214, 219, and 220.

Because I am "conceding" here (i.e., recognising and noting facts that go "against" the strength of my hypothesis), it is not necessary for our purposes to justify this recognition. The claims that substantiate my hypothesis are the ones that are justified (in the next section).

4.3 The Properties that Parapsychology (Mostly) Does Not Have

In light of its lack of consistent replication (addressed below), parapsychology follows an unorthodox and irrational general method: "To insist that in certain special cases where we have no consistent repeatability, and also have no theoretical understanding of a possible connection, we must nevertheless be able to *discover* some theoretical underpinning shows a procedure remarkably divergent from our usual approach to unknown areas" in science (Pletcher, 1990, p. 213) (lack of properties 65 and 103–104).

Parapsychology lacks corroborated explanations made with respect to laws, and cannot predict specifically and successfully. "[N]ot a single lawful interrelationship appears to have been established" in parapsychology, and "each reported investigation yields a result that has little relationship to any of the others" (Hansel, 1989, p. 267). "A real explanation would cite the [...] mechanisms involved, identify the empirical facts that reflect them, appeal to the laws that govern them, and then deduce exactly the surprising effect observed" (lack of properties 45 and 181–182); one of parapsychology's most important deficits is that type of "uniquely successful explanation in terms of some detailed [...] theory" (Churchland, 1987, pp. 234–5) (lack of property 48). And it suffers from a lack of fecundity/fruitfulness of predictions and of successful predictions. "[W]e cannot predict *when* or *under what conditions* above-chance calls will be made" (Kurtz, 1981, p. 12, emphasis in original) (lack of property 82).

In keeping with our properties 27, 210, and 215, the investigation of anomalies is a scientific virtue. Therefore, parapsychology should not be considered pseudoscientific because it investigates anomalies. But not even the existence of those anomalies has been established to the general satisfaction of the scientific community after more than a century and a half of research (explanations of such alleged anomalies that invoke the paranormal are a fortiori even less scientific, because they additionally violate the principles of parsimony and bulk external consistency) (the principle of parsimony in explanation is, at best, inconsistently applied in parapsychology (Pletcher, 1990, p. 212)) (lack of properties 81 and 190). Parapsychology has been unsuccessful in obtaining sufficient evidence: there are few positive results confirming the existence of such occurrences (Flew, 1990, p. 215) (lack of property 100). Because of this, parapsychology appears to give excessive weight to confirmation over falsifiability or over falsification (lack of property 124). The experiences (i.e., observations or experiments) allegedly/apparently confirming it could not be successfully and consistently repeated/replicated by different (already established as) scientific investigative bodies and researchers that attempted their replication (ibid.; Bunge, 1967; Beloff, 1973, 1977, p. 759, 1978, 1984; Flew, 1978; Kurtz, 1981, pp. 12–3, 16, 18, 21–2; Blackmore, 1983; Palmer, 1985; Alcock, 1987a, pp. 553, 563, 565; Churchland, 1987, pp. 236–7; Spanos & de Groot, 1987; Hansel, 1989, pp. 265, 268–70; Grim, 1990, pp. 183–4; Reinsel, 1990, p. 195; Pletcher, 1990, pp. 205, 212–3; Flew, 1990, pp. 223–4; Rhine, 1990, pp. 256–7; Sommer, 2014, p. 41). This is the case even though "[r]eplicability does not necessarily mean that a finding must be reproducible on demand" (Rao & Palmer, 1987a, p. 544). Allegedly successful parapsychological experiments could not be successfully and consistently repeated/replicated "independently of the investigators in charge" (lack of property 57) (Hansel, 1989, p. 271). "Some parapsychologists seem consistently to obtain the results they desire whereas others are unable to find significant departures from chance (Palmer, 1985)" (Alcock, 1987a, p. 561). Their allegedly procedurally valid studies give inconsistent results (lack of property 211). "[T]he fact that the outcomes of psi experiments seem to be sensitive [...] to the identity of the experimenter or principal investigator is a legitimate cause for concern. This 'experimenter effect' in parapsychology has long been recognized and extensively discussed within the field (e.g., Kennedy & Taddonio, 1976; White, 1976a, 1976b, 1977); even some strong proponents of psi have had trouble obtaining positive results in their experiments" (Rao & Palmer, 1987a, p. 548). "It has been observed from the early days of [...] research into psi that the experimenter is a relevant variable. It is now well known that some researchers are more successful than others in obtaining significant psi scores" (Rao, 2001, p. 16).

Test subjects do not regularly demonstrate "the ability to obtain high scores" when extra safeguards and precautions are taken (Gardner, 1987, p. 587; cf. Hansel, 1989, p. 155) or when the test procedures are "completely mechanized" (ibid., p. 268). And "one can seriously challenge the claim that practitioners of normal science do not give, or have not given, parapsychology its day in court" (Alcock, 1987a, pp. 563–4; Hacking, 1993, pp. 589–91). "A great deal of experimental work has failed to produce a clear case for the existence of ESP"; a "great deal of time, effort, and money has been expended over a period of more than a hundred years, but an acceptable demonstration of the existence of extrasensory perception or psychokinesis has not been provided" (Hansel, 1989, pp. 264, 272). "For over a century, there has been careful and deliberate investigation of psi phenomena by people trained in the methods of science. [...] [M]uch of this research has been laboratory-based and carried out in university settings" (Alcock, 1987a, p. 553). If mere statistical (significance in) replication of successful parapsychological experiments existed (as claimed e.g. in Radin et al., 2006), then we should examine the individual tests for flaws, take into account possible weaknesses due to selection of the material published (Hansel, 1989, pp. 269–70), and look for normal or ordinary causes of the effect. "Minuscule interactions and minuscule changes in a physical system can be due to an enormous variety of possible, ordinary causes, and we have no inventory of all the experimental protocols and descriptions for some mundane explanation of the minuscule" (Glymour, 1987, p. 590). "Even very small percentages of errors, hidden artifacts, dishonesty, or nonindependence, let alone publication bias, would make the proportion higher than that expected by chance. It is thus doubtful that there exist in psychology many null hypotheses that cannot be brought to the point of rejection by some form of statistical meta-analysis" (Navon, 1987, p. 599). Therefore, parapsychology does not apply Ockham's razor (lack of property 164). "[P]arapsychologists are no nearer finding [...] a demonstration than they were a hundred years ago" (Hansel, 1989, p. 271) (lack of properties 100, 183–184, and 213).

Parapsychology is isolated and non-externally-systematic in that it "is not a component of the system of the sciences" (Bunge, 1987, p. 576; Alcock, 1987a, p. 562; Bunge, 1991, p. 135) (lack of properties 57 and 189). The simple referencing or citation of clearly scientific works (e.g., articles in mainstream science journals) in the parapsychological field (cf. Mousseau, 2003) does not per se demonstrate that parapsychology is a component of the system of the sciences.

The alleged anomalies studied by parapsychology, or at least any paranormal explanation of them, are externally inconsistent in bulk with science (McClenon, 1982; Rao, 1983): they clash "with either well-established scientific theories or fertile philosophical principles" (Bunge, 1987, p. 576, 1991, p. 147) and lie "outside the scientific worldview as we know it today" (Alcock, 1987b, p. 627) (lack of property 190). These "interactions [are called] anomalous because they appear to exceed somehow the capacities of the sensory and motor systems as these are presently understood", and in this way they "appear to deviate from commonly accepted notions regarding the limitations of sensory/motor capacities" (Rao & Palmer, 1987a, p. 539, 1987b, p. 618), as they "involve some radically different relationship between consciousness and the physical world than that held to be possible by contemporary science" (Alcock, 1987a, p. 556), implying perception (or "prehension"), or influence, of/over "things that are not within the organism's ordinary [...] boundaries" of perception (or "prehension") (Carpenter, 2012, p. 287). "[I]f [paranormalist] parapsychologists are right about psi, then the well-tested theories of [...] neurologists [and neurophysiologists] are wrong" (Alcock, 1981, 1987a, pp. 562, 565). The simple referencing or citation of clearly scientific works (e.g., articles in mainstream science journals) in the parapsychological field (cf. Mousseau, 2003) does not per se amount to overlap with other scientific fields. If human ESP existed, then the "prevailing ideas in psychology and sensory physiology" (Hansel, 1989, p. 273), "plus the philosophy inhering in them" (Bunge, 1991, p. 136), would be essentially mistaken, because "ESP is at variance with what is known about Perception and about the relationship between psychological phenomena, the nervous system and the brain" (Hansel, 1989, p. 298) and it conflicts "with well-established principles" (ibid., p. 13). "If a 'paranormal' theory of psi were someday to be confirmed, the practical consequence would be a redefinition of 'normal' science" (Rao & Palmer, 1987a, p. 540). Alleged paranormal parapsychological occurrences/facts are considered "paranormal" precisely because of this characteristic. This is why the claims of paranormal parapsychology are extraordinary, which requires extraordinarily strong evidence. The failure for over a century and a half to provide confirmed parapsychological observations "indicates that prevailing ideas in psychology and sensory physiology are essentially correct" (Hansel, 1989, p. 273).

Parapsychology is also externally inconsistent in that the idea is defended within parapsychology that it is a research field in which the very nature of the effects studied allows for a uniquely legitimate role of the experimenter/observer effect or of the sheep-goat effect[4] (e.g., Rao, 2001, pp. 16–8: "A strong case can be made [...] that experimenter effects observed in parapsychological research are genuine psi effects"). These are sometimes interpreted to be "from the psi influence of the experimenter" (Krippner, 1978; Palmer, 1985, 1986; quote: Alcock, 1987a, p. 561; Rao & Palmer, 1990, p. 388). Therefore, parapsychology applies its problem-solving strategies with auxiliary hypotheses that can be "tested" only in their applications, it has *built-in subterfuge*, and its auxiliary hypotheses are not independently testable (lack of properties 125 and 185–187).

The paranormal hypothesis in parapsychology (also known as the psi hypothesis) becomes unfalsifiable if a paranormal ("psi") experimenter/observer effect, or a paranormal ("psi") sheep-goat effect, or uncheckable mental or emotional states of the test subjects, or a paranormal "interpersonal context in the moment" (Carpenter, 2012, p. 289) are invoked to justify that results approximate chance or mean chance expectation results when agnostics or sceptics are involved in any way in the tests or experiments (including test design and supervision of test execution), or when extra safeguards and precautions are taken during the tests or experiments. This is because such justification would "make the psi hypothesis unfalsifiable by providing ways to explain away null results and nonreplications" (Alcock, 1987a, pp. 553, 561–2).

Parapsychology's inconsistency with currently accepted scientific ideas is even stronger, because it involves physics, if it is claimed that ESP or psychokinesis, while being physical or natural, are independent of distance (i.e., do not decline with it, violating the inverse-square law); or violate the law of conservation of energy, or the inverse-square law (Kurtz, 1981, pp. 13–4, 19–20; Gardner, 1981, pp. 62–3; cf. Bunge, 1991, p. 136; e.g., Murphy, 1961; and: "Psi results seem to violate expectations derived from generally accepted physical theory" (Rao & Palmer, 1987a, p. 550; see also Rao, 2001, pp. 20–2)). Also, in ESP "[i]t would

[4] The sheep-goat effect is the phenomenon that tests in which the participants are believers in the reality of ESP or psychokinesis, or in the possibility of any of them under a given experimental setting ("sheep"), tend to yield results that are more favourable for the relevant belief that tests performed by sceptics ("goats"). According to a former president of the Parapsychological Association, James Carpenter, it is a "relatively robust" effect (2012, p. 281).

seem that [...] the subject has access to information under conditions that [...] do not permit any known physical energy transmission from the target" (Rao, 2001, p. 21).

Parapsychological research has mostly been not theoretically-directed (lack of property 111).

The ways in which parapsychologists interpret their results show assumptions of explanatory frameworks that vary "in a far too ad hoc way to be satisfactory". "It would appear that those characteristics are assigned to psi that are necessary to explain deviations from expectations" (Pletcher, 1990, p. 211). "[I]t is standard practice in parapsychology to take one pattern of data as evidence for psi in one experiment, then to disregard its absence and take some other pattern as evidence for psi in another experiment" (Alcock, 1987a, p. 558). "[P]arapsychologists can slip from one ad hoc explanation to other" (quote: Kurtz, 1981, p. 19; Alcock, 1987a, pp. 556, 562) (lack of properties 9–10).

Parapsychology has been ostracised by the international psychological community since 1905 (Benjamin & Baker, 2012, pp. 3–4). Parapsychology is not scientifically peer-reviewed in scientific publications (parapsychologists do perform peer-review of their peers; and parapsychological propositions can be, and are sometimes, examined by science). Parapsychology is not subject to scientific peer review (lack of property 128) and avoids scientific peer review (lack of property 208) by having its own publications.

Regarding all the properties in the demarcatory list not mentioned in this chapter, it is unclear whether parapsychology mostly has them. Non-consensual proposals of mechanisms, hypotheses, and theories of varied degrees of vagueness and in diverse stages of development have been put forward for paranormally explaining parapsychology's alleged anomalies. "Overviews of theories in experimental parapsychology are Rao (1977, 1978) and Stokes (1987). More recent theoretical proposals are [...] Atmanspacher et al. (2002) and Carpenter (2012)" (Sommer, 2014, p. 41).

4.4 Chapter Conclusions

Parapsychology lacks at least 28 of the properties of scientificity featured in the demarcatory list, while possessing at least 18 of the specific ones. This yields a score of −10 for parapsychology. This (possibly provisional) tally is strongly against the scientificity of parapsychology, and is the best explanation for the fact that parapsychology is commonly deemed non-scientific. In this case, the demarcatory list of criteria again does a good job of capturing the consensus about which disciplines are sciences/scientific.

References

Alcock, J. E. (1981). *Parapsychology: Science or magic?* Pergamon Press.

Alcock, J. E. (1987a). Parapsychology: Science of the anomalous or search for the soul? *Behavioral and Brain Sciences, 10*(4), 553–643.

Alcock, J. E. (1987b). A to-do about dualism or a duel about data? *Behavioral and Brain Sciences, 10*(4), 627–643.

Atmanspacher, H., Römer, H., & Walach, H. (2002). Weak quantum theory: Complementarity and entanglement in physics and beyond. *Foundations of Physics, 32*, 379–406.

Beloff, J. (1973). *Psychological sciences*. Crosby Lockwood Staples.

Beloff, J. (1977). Parapsychology and philosophy. In B. B. Wolman (Ed.), *Handbook of parapsychology*. Van Nostrand Reinhold.

Beloff, J. (1978). Why parapsychology is still on trial. *Human Nature, 1*(12), 68–74.

Beloff, J. (1984). Research strategies for dealing with unstable phenomena. *Parapsychology Review, 1*, 1–7.

Benjamin, L. T., Jr., & Baker, D. B. (2012). The internationalization of psychology: A history. In D. B. Baker (Ed.), *Oxford handbook of the history of psychology*. Oxford University Press.

Blackmore, S. (1983). "Unrepeatability: Parapsychology's only finding", paper presented at the Parapsychology Foundation conference, San Antonio, Texas, October of 1983.

Bunge, M. A. (1967). *Philosophy of science: From problem to theory*. Springer.

Bunge, M. A. (1987). Why parapsychology cannot become a science. *Behavioral and Brain Sciences, 10*(4), 576–643.

Bunge, M. A. (1991). A skeptic's beliefs and disbeliefs. *New Ideas in Psychology, 9*(2), 131–149.

Carpenter, J. C. (2012). *First sight. ESP and parapsychology in everyday life*. Rowman & Littlefield.

Churchland, P. M. (1987). How parapsychology could become a science. *Inquiry: An Interdisciplinary Journal of Philosophy, 30*(3), 227–239.

Flew, A. (1978). Parapsychology revised: Laws, miracles, and repeatability. In J. Ludwig (Ed.), *Philosophy and parapsychology* (pp. 263–269). Prometheus Books.

Flew, A. (1990). Parapsychology: Science or pseudoscience? In P. Grim (Ed.), *Philosophy of science and the occult* (2nd ed., pp. 214–231). State University of New York Press.

Gardner, M. (1981). Einstein and ESP. In K. Frazier (Ed.), *Paranormal borderlands of science* (pp. 60–65). Prometheus Books.

Gardner, M. (1987). Evidence of the paranormal: A skeptic's reactions. *Behavioral and Brain Sciences, 10*, 587–588.

Glymour, C. (1987). ESP and the big stuff. *Behavioral and Brain Sciences, 10*, 590.

Grim, P. (1990). Parapsychology. Introduction. In P. Grim (Ed.), *Philosophy of science and the occult* (2nd ed., pp. 183–186). State University of New York Press.

Hacking, I. (1993). Some reasons for not taking parapsychology very seriously. *Dialogue, XXXII*, 587–594.

Hansel, C. E. M. (1989). *The search for psychic power: ESP and parapsychology revisited*. Buffalo, New York: Prometheus Books (this is a "third edition" under a novel title, as it is a revised edition of Hansel, C. E. M. (1980), *ESP and parapsychology: A critical re-evaluation*. Buffalo, New York: Prometheus Books, which is in turn a revised edition of Hansel, C. E. M. (1966), *ESP: A scientific evaluation*. New York: Charles Scribner's Sons).

Kennedy, J. E., & Taddonio, J. (1976). Experimenter effects in parapsychological research. *Journal of Parapsychology, 40*, 1–33.

Krippner, S. (Ed.). (1978). *Advances in parapsychological research, Vol. 2: Extrasensory perception*. Plenum Press.

Kurtz, P. (1981). Is parapsychology a science? In K. Frazier (Ed.), *Paranormal borderlands of science* (pp. 5–23). Prometheus Books.

McClenon, J. (1982). A survey of elite scientists: Their attitudes toward ESP and parapsychology. *Journal of Parapsychology, 46*, 127–152.

Mousseau, M.-C. (2003). Parapsychology: Science or pseudo-science? *Journal of Scientific Exploration, 17*, 271–282.

Murphy, G. (1961). *The challenge of psychical research*. Harper & Row.

Navon, D. (1987). On rustles, wolf interpretations, and other wild speculations. *Behavioral and Brain Sciences, 10*, 599–600.

Palmer, J. (1985). Psi research in the 1980s. *Parapsychology Review, 16*, 1–4.

Palmer, J. (1986). ESP research findings: The process approach. In H. L. Edge, R. L. Morris, J. H. Rush, & J. Palmer (Eds.), *Foundations of parapsychology*. Routledge & Kegan Paul.

Pletcher, G. K. (1990). Coincidence and explanation. In P. Grim (Ed.), *Philosophy of science and the occult* (2nd ed., pp. 205–213). State University of New York Press.

Radin, D., Nelson, R., Dobyns, Y., & Houtkooper, J. M. (2006). Reexamining psychokinesis: Comment on Bösch, Steinkamp, and Boller (2006). *Psychological Bulletin, 132*(4), 529–532.

Rao, K. R. (1977). On the nature of psi: An examination of some attempts to explain ESP and PK. *Journal of Parapsychology, 41*, 294–351.

Rao, K. R. (1978). Theories of psi. In S. Krippner (Ed.), *Advances in parapsychological research* (Vol. 2, pp. 245–295). Plenum.

Rao, K. R. (1983). From proof to acceptance. *Parapsychology Review, 14*, 1–3.

Rao, K. R. (2001). Introduction: Reality, replicability, and lawfulness of psi. In K. R. Rao (Ed.), *Basic research in parapsychology* (2nd ed., pp. 3–37). McFarland.

Rao, K. R., & Palmer, J. (1987a). The anomaly called psi: Recent research and criticism. *Behavioral and Brain Sciences, 10*, 539–551.

Rao, K. R., & Palmer, J. (1987b). Where lies the bias? *Behavioral and Brain Sciences, 10*, 618–627.

Rao, K. R., & Palmer, J. (1990). Researching data and searching for theory. *Behavioral and Brain Sciences, 13*, 387–389.

Reinsel, R. (1990). Parapsychology: An empirical science. In P. Grim (Ed.), *Philosophy of science and the occult* (2nd ed., pp. 187–204). State University of New York Press.

Rhine, J. B. (1990). Second report on a case of experimenter fraud. In P. Grim (Ed.), *Philosophy of science and the occult* (2nd ed., pp. 253–263). State University of New York Press.

Sommer, A. (2014). Psychical research in the history and philosophy of science. An introduction and review. *Studies in History and Philosophy of Biological and Biomedical Sciences, 48*, 38–45.

Spanos, N. P., & de Groot, H. (1987). Psi: Repeatability, falsifiability, and science. *Behavioral and Brain Sciences, 10*, 609–610.

Stokes, D. M. (1987). Theoretical parapsychology. In S. Krippner (Ed.), *Advances in parapsychological research* (Vol. 5, pp. 77–189). McFarland.

White, R. A. (1976a). The influence of persons other than the experimenter on the subjects' scores in psi experiments. *Journal of the American Society for Psychical Research, 70*, 133–166.

White, R. A. (1976b). The limits of experimenter influence on psi test results: Can any be set? *Journal of the American Society for Psychical Research, 70*, 333–369.

White, R. A. (1977). The influence of experimenter motivation, attitudes, and methods of handling subjects in psi test results. In B. B. Wolman (Ed.), *Handbook of parapsychology* (pp. 273–301). Van Nostrand Reinhold.

Chapter 5
The Properties of Creation "Science"

Abstract This chapter provides a brief definition and a description of/introduction to creation "science". It then meta-analytically shows that the demarcation criteria and the scoring method proposed in Chap. 2 demarcate creation "science" as non-scientific, in agreement with the current philosophical and scientific consensus.

Keywords Creation "Science" · Intelligent design · Non-science · Pseudoscience · Demarcation · Scientificity

5.1 Introduction

This chapter shows that the demarcation criteria demarcate creation "science" as non-scientific, in agreement with the current philosophical and scientific consensus. This section defines creation "science". Section 5.2 explains why certain properties of creation "science" do not count against its scientificity. Section 5.3 lists properties of scientificity that creation "science" has. Section 5.4 lists properties of scientificity that creation "science" lacks. As the conclusion of this chapter, Sect. 5.5 argues that the demarcation criteria listed in Sect. 2.2 demarcate creation "science" as non-scientific (that creation "science" does not have enough of the properties of scientificity, and therefore should be deemed non-scientific), and that the fact that creation "science" is commonly deemed non-scientific is explained by the fact that it does not have enough of the properties of science listed in Chap. 2.

I take the properties of science listed in Sect. 2.2, and show that creation "science" lacks many of them. This provides: a) a rational reconstruction (and therefore an explanation) of why creation "science" is non-scientific, b) further evidence of the extensional adequacy of this work, c) further evidence of this work's explanatory power, and d) further evidence of the fact that the proposed set of criteria demarcates science appropriately.

Creation "science", also known as "scientific" creationism, is a CF "which posits that [sufficient] scientific evidence exists to prove that the universe and [the] living things [in it] were specially created in their present form" (or in certain original

D. Fernandez-Beanato, *A Working Scientific Demarcation*, Synthese Library 519, https://doi.org/10.1007/978-3-031-99005-2_5

basic types) by an intelligent designer (National Academy of Sciences, 1999). This rules out biological macro-evolution. Here, "specially" is a technical theological term which means "supernaturally", i.e., in a manner above and beyond the order of nature, "actively steering [...] by way of miraculous intervention" (Plantinga, 1991, p. 22; Mc Mullin, 1993, pp. 299–300, quote: 304, 329; Gish, 1995, p. 42; National Academy of Sciences, 1999, p. 7; Sober, 2008, p. 112). Therefore, the creation of a chair by a carpenter, or even the creation of a new biological species via genetic engineering, would not be instances of "special" creation. That is why the creator entity referred to in creation "science" is a god, or a god-like being.[1] This creationist hypothesis is not necessarily religious, as it does not require its believers to adopt any devotional attitude or particular behaviours towards the creator entity. The creator in question does not have to be good, or have any connection to any particular religious tradition.

There is evidence (empirical support) for "intelligent design" in nature, in the sense that there are undisputedly true propositions about nature that are entailed by "intelligent design" theories. Consequently, facts in nature support and confirm the intelligent design theories because the theories entail said facts. But this is simply because those theories were concocted to explain facts in nature, and therefore trivially entail facts in nature (cf. Mc Mullin, 1993, p. 305).

Creation "science" is often considered to be a prime example of a non-scientific and pseudoscientific CF (cf. Mc Mullin, 1993, p. ix; Boudry, 2011, p. 4) by both philosophers and scientists.

5.2 The Problem with Methodological Naturalism

The American National Academy of Sciences (1999, p. 7) has indicated that the claims of creation "science" do "not qualify as scientific hypotheses" in part because they "do not refer to natural causes and cannot be subject to meaningful tests". But the proposition that the claims of creation "science" "cannot be subject to meaningful tests" is false: science has geological, biological, and other types of evidence that falsify those claims (as, contradictorily, the very same National Academy of Sciences document referenced makes very clear) (cf. Laudan, 1982, pp. 16–7).

Furthermore, reference to non-natural causes is not disqualifying as long as traces of their interactions with other spheres of reality can be found (Plantinga, 1991, p. 27; Boudry, 2011, 2017; Boudry & Leuridan, 2011). Hansson (2013, p. 64) proposed that the subject matter of scientific knowledge must be exclusively about

[1] Therefore, creation "science" is different from the "ancient astronauts" hypothesis that non-supernatural extraterrestrial beings created humans, or influenced the process of hominisation, by using genetic engineering. Some authors contend that because of the Third Law of Clarke (which states that "Any sufficiently advanced technology is indistinguishable from magic"), it is impossible for us to reliably discern here between gods and non-divine beings endowed with what for us would be god-like powers.

nature, ourselves, our societies, and our physical and thought constructions. But this leaves unanswered the question of why a science's object of knowledge could not be something else, for example the supernatural. A partially correct usual answer is that this very concept is badly delineated in most cases. As this argument goes, if a phenomenon or a being once considered to be supernatural is finally confirmed to exist, there is no reason why it could not be integrated as a part of a previously badly understood natural world. For example, while most religious traditions regard gods as supernatural entities, many Buddhists and Jains consider the gods in their pantheons to be natural beings, parts of an uncreated all-natural universe. But even if we accept this objection, a universal creator god would certainly be a clear case of an entity falling under the category of the supernatural (meaning "above the natural world") (cf. Clarke, 2009, p. 130), unless we regarded everything in existence, or everything interacting with the natural world, as being natural by definition (Clarke, 2009, pp. 138–9).

When assessing statements about gods, we should avoid the interpretation that most logical empiricists gave to their empiricist criterion of meaning, which wrongly ruled out too many statements as meaningless, including propositions about gods. Atheists and agnostics have often employed it, but it seems undeniable that millions of people hold cognitively meaningful, sometimes quite specific beliefs about gods (Hume also knew this much). Quine summarised the convenience of taking an intermediate approach between these two positions, by stating: "We need no longer labour under the delusion that the meaningfulness of a statement containing a singular term presupposes an entity named by that term", in a nutshell, there is meaning without being, or, as he put it, "A singular term need not name to be significant" (Quine, 1961, p. 9).

A commonplace in secular academic thought is that the problem with any theology that claims to have knowledge acquired via revelation is that inter-subjective acceptance of that alleged knowledge is not achieved, for the most part, by the communication of rationally- and/or empirically-justified statements. Rather, it is predominantly instilled in early education and/or transmitted or adopted through emotional means or mechanisms—or so the argument goes. But what about non-revealed—transcendental or natural—theologies, which are supposed to advance rational and empirical arguments, respectively? The whole of theology—a discipline dealing with some clearly supernatural proposed entities—, even in its non-revealed variety, is sometimes excluded from SBS because of methodological naturalism, i.e., the doctrine that science is inherently limited to dealing with the natural world and formulating natural explanations of it through natural means. This clearly seems to be a *petitio principii* or a special plea. This is how Boudry highlights the empirical side of the argument for rejecting the demands of methodological naturalism:

> [N]arrowing down the scope of science by excluding all supernatural claims from its purview is unfeasible and historically inaccurate, given that many such claims have in fact been subjected to empirical investigations (e.g., the healing power of intercessory prayer [...], communication with angels). Upon any nontrivial definition of the term "supernatural," I

> see no sound reason why supernatural phenomena would be intrinsically beyond the pale of science. (Boudry, 2013b, pp. 85–6)

And why he and Fishman and Boudry (2013) argue that

> science presupposes neither MN [methodological naturalism] nor ON [ontological naturalism] and [...] science can indeed investigate supernatural hypotheses via standard methodological approaches used to evaluate any 'non- supernatural' claim. Science, at least ideally, is committed to the pursuit of truth about the nature of reality, whatever it may be, and hence cannot exclude the existence of the supernatural *a priori*, be it on methodological or metaphysical ground, without artificially limiting its scope and power. (Fishman & Boudry, 2013, p. 921)

Ideally, from a scientific point of view, all types of claimed empirical or mental phenomena should be researched to ascertain the merits of the claims (cf. Fishman, 2009). And *prima facie*, it is difficult to indicate a reason why all proposed theological explanations of phenomena should be deemed non-scientific *a priori*.

And it is not difficult to *imagine* clear evidence of a miraculous phenomenon (Fales, 2013, p. 251). To illustrate: in September of 1995, a Hindu devotee in New Delhi, India, observed a *murti* (statue) of the elephant-faced god Gaṇesha "drinking" milk from a spoon. The liquid slowly disappeared from the piece of cutlery when it was applied to the *murti*'s trunk. Other worshippers who offered milk to various other gods' statues successfully replicated the phenomenon, which made it to the news and quickly spread to the United Kingdom, Canada, the United Arab Emirates, Nepal, and other countries. Numerous faithful witnesses did not only see the phenomenon, but also, unlike in earlier times, were able to record it on video (which ruled out it being a case of mass delusional hysteria). Thus, the occurrence of the phenomenon was established beyond any reasonable doubt, so as to merit an explanation. The Hindu theological explanation, espoused by the World Hindu Council, was that Gaṇesha was physically showing his acceptance of the lactic oblations. An alternative explanation was that the milk was being pulled up and out of the containers by capillary motion, which is a widely known natural process. The latter solution was confirmed, among other evidence, when the phenomenon was replicated using a Mickey Mouse figurine, a possibility implicitly prohibited by the theological hypothesis, since the Disney rodent is clearly not a deity. Even though in this case everything turned out to be consistent with a metaphysical naturalist conception of the world, my moral for this story is that, had the phenomenon been conclusively proven to occur exclusively through the mouths of statues in the shapes of Hindu idols, naturalists would have had a lot more to explain. Other good examples of imagined evidence of miraculous phenomena are given by Dembski (2000, p. 256) and Boudry (2011, pp. 21, 89–90).

But is there a necessity for any theological claim to have direct empirical consequences in order for it to be within the scope of SBS? Some branches of SBS, such as metaphysics and logic, are not so much in touch with the empirical realm in their investigations, but they still intend to talk about what *is* there, and appear to do a reasonably good job at it. To begin with, metaphysics succeeds in formulating synthetic statements that are meaningful even if not empirically testable, as Popper pointed out against logical positivism (cf. Agassi, 1964) (incidentally, this last claim can also be made of theology). Also, if we accept the classical analytic-synthetic distinction, logic and mathematics formulate meaningful statements that are not synthetic; therefore they do not require empirical tests in order to demonstrate their truthfulness.

So it appears that SBS's subject matter is *what is, what was, what will be* and *what could be*—as in Aristotle's (1961 [4th century BCE]) definition of what is true and what is false (*Metaphysics*, IV.7)—, rather than what is, what was, what will be and what could be exclusively *in the natural world*. As a thought experiment, we can consider the case of a god that did not, is not and will not interfere with the natural universe he once created; only first-cause considerations about that universe's origin could potentially reveal that god's existence to the natural sentient beings he created. We could even imagine a god that never created a natural universe; a natural theology about him could not exist in the absence of a natural universe, but a transcendental theology would still be possible, even if only in its own mind, or the minds of other beings.

So, if the problem with creation "science" is neither that it appeals to supernatural causes nor that it is not testable, the problem must lie somewhere else. Let us see how creation "science" stands regarding the possession of our scientificity properties.

5.3 The Properties of Creation "science"

We can concede that the best creation "science" satisfies the following properties: 15, 29, 31, 34, 51–52, 54, 68, 85, 91, 176, and 188.

Because I am "conceding" here (i.e., recognising and noting facts that go "against" the strength of my hypothesis), it is not necessary for our purposes to justify this recognition. The claims that substantiate my hypothesis are the ones that are justified (in the next section).

5.4 The Properties that Creation "science" (Mostly) Does Not Have

As it happens with genethliac astrology, different versions of creation "science" are unfalsifiable (for example, any version claiming that everything is exactly as the Creator wanted it to be (cf. Kitcher, 1981, pp. 527–8),[2] which is a way of "insulating the design argument against empirical objections" (Boudry & Leuridan, 2011, p. 574)) or, alternatively, falsified (by the geological and biological evolutionary

[2] More subtle problems with its testability are detailed in Sober, 2008, pp. 109–88. For example, we don't know how to suitably choose auxiliary hypotheses for this theory.

records) (either lack of properties 52 and 54, or of property 184). In the second case, "the design argument [...] has simply collapsed under the weight of evidence and has been outcompeted by evolutionary theory" (Boudry & Leuridan, 2011, pp. 559, 571). There is no empirical competition between creation "science" theories and Darwinian theory (here, "Darwinian" is used in a broad sense) (Sober, 2008, p. 184) (for an overview, see Dawkins, 2009; Coyne, 2009). Creation "science" is also dogmatic in its "refusal to face [the] facts and debate rationally" (quote: Bunge, 2013, p. 24) (lack of property 117).

Creation "science" accomplishes only a spurious unification or consilience, because of overfitting and under-specification (cf. Mc Mullin, 1993, pp. 317, 319) (lack of properties 46 and 81): "[a]s Kitcher has pointed out, [...] in such an explanatory pattern the "nonlogical vocabulary which remains is idling" (1981, [p.] 528). The pattern does not impose constraints on the sentences that can be derived by using it, and thus it is able to accommodate any observation whatsoever. In Sober's vocabulary, the only reason for adopting the assumption that God really wants a specific fact O_i to be the case depends on looking at O_i and nothing else." (Boudry & Leuridan, 2011, pp. 563–4). The auxiliary hypotheses/assumptions required for inferring creation "science" conclusions are not independently established, "we have no independent knowledge of the[m]" (Sober, 2008, pp. 141–6, 154) (lack of properties 58, 125, and 187). "Thus, in reality the design theorist simply posits a new divine disposition for each and every observation instead of a single unifying explanation." (Boudry & Leuridan, 2011, p. 564). In this way, she "treats each observation as an isolated phenomenon in its own right" (Sober, 2008, p. 184). This constitutes extreme disunification, suffered by all creation "science" models proposed so far (Sober, 2008, pp. 182–4). To avoid it, the model or theory must have fewer parameters than observations to explain, and in order to accomplish this, one "need[s] to collect different observations together and view them as consequences of a single plan that the designer(s) had in mind" (Sober, 2008, p. 182). Creation "science" lacks predictive power: it does not generate novel predictions, because "[i]t is designed to yield the known observations and nothing more" (Boudry, 2011, pp. 28–9, 87–93, 235; quote: Boudry & Leuridan, 2011, pp. 563; cf. p. 575; Worrall, 2004, p. 68) (lack of properties 45, 82, and 184). "From a likelihood point of view, the problem with the design hypothesis is that we have no independently justified auxiliary propositions that permit the hypothesis to make predictions about the observations" (Sober, 2008, p. 181). "[T]he theory is too vague to allow for specific predictions [(Mc Mullin, 1993, pp. 318–9)] and to achieve any form of genuine explanatory unification" (Boudry, 2011, p. 235, 2013, p. 86). "[T]he doctrine of special creation is [...] empty; it allows each observation to be described *post hoc*, and that is all. What is wrong with such 'theories' is that they do not make predictions" (Sober, 2008, p. 183). "The problem with the hypothesis of intelligent design is not that it makes inaccurate predictions but that it doesn't predict much of anything. Rather, the design hypothesis merely allows our observations – whatever they turn out to be – to be folded inside a simple formula", and that that formula's schema could include any possible observation (ibid., p. 154) (lack of properties 42, 48, 82, and 105). Creation "science" yields nothing beyond observations and the idle proposal of a Creator (ibid., pp. 90, 182).

Creation "science" is not specific enough in its descriptions, predictions, and explanations (there is no fleshing out of its design hypothesis). Supporters of creation "science" are completely unwilling to, or incapable of, sufficiently fleshing out their design hypothesis (Boudry, 2011, pp. 141, 235; Boudry & Leuridan, 2011, p. 574). Creation "science" does not add sufficient "empirical substance to the theory" (cf. National Academy of Sciences, 1999, p. 1; Pennock, 1999, p. 249; quote: Boudry & Leuridan, 2011, p. 574). Creation "science" provides us with no clue as to the mechanism of creation or the Creator's modus operandi, other than the fact that it "created all living beings in the world separately" and maybe certain proposed timings (Boudry, 2011; Boudry & Leuridan, 2011 (quote: pp. 570–1)) (lack of properties 46 and 179).

Creation "science" is isolated and non-externally-systematic in that it "is not a component of the system of the sciences" (quoted phrase: Bunge, 1987, p. 576) (lack of properties 57 and 189). "Documentation offered in support of these claims is typically limited to the special publications of their advocates" (National Academy of Sciences, 1999, p. 25). This means that creation "science" is not scientifically peer-reviewed in scientific publications (creation "scientists" do perform peer-review of their peers; and creation "science" propositions can be, and are, examined by science). Creation "science" is not subject to scientific peer review (lack of property 128) and avoids scientific peer review (lack of property 208) by having its own publications.

Creation "science" is theoretically, methodically, and conceptually externally inconsistent (Reisch, 1998, pp. 334–9) (lack of property 190). A creation "science" "conception of species would treat them as independent metaphysical kinds, and would perhaps dovetail with creationist theories specifying the limits within which species can split or evolve" (ibid., pp. 338–9).

One of the main arguments for creation "science" is that of "irreducible complexity" (i.e., the idea that some complex biological systems could not have evolved by a chain of minute modifications to functional systems via natural selection, because their component parts would neither provide any selective advantage nor perform any beneficial function in isolation). This argument, as used in creation "science", suffers from inherent conceptual equivocations (see Dunkelberg, 2003; Boudry, 2011; Boudry & Leuridan, 2011) (lack of property 129); or, if it is not equivocal, then it is subjected to the "wine-bottle problem" about how to decompose systems into parts (Sober, 2008, pp. 135–6, 155). William Paley's (1809) concepts of usefulness and intolerance are antecedents of the concept of irreducible complexity (Sober, 2008, p. 155).

Because of the geological and biological evolutionary evidence against the falsifiable versions of creation "science", their supporters resort to "gerrymandering hypotheses by inventing ad hoc auxiliaries to fit the data" (Boudry & Leuridan, 2011, p. 86) (lack of properties 10 and 60).

Modern creation "science" "advocates have made little progress since Paley" (ibid., p. 89) (lack of property 84).

Regarding all the properties in Sect. 2.2 not mentioned in this chapter, it is unclear whether creation "science" mostly has them.

5.5 Chapter Conclusions

Creation "science" lacks at least 22 of the properties of scientificity featured in the demarcatory list, while having at least 4 of the specific ones. This yields a score of −18 for creation "science". This (possibly provisional) tally is strongly against the scientificity of creation "science", and is the best explanation for the fact that creation "science" is commonly deemed non-scientific. This, once again, demarcates in a manner that is equivalent to that of the current international philosophical and scientific consensus regarding the extension of the term "science".

References

Agassi, J. (1964). The nature of scientific problems and their roots in metaphysics. In M. A. Bunge (Ed.), *The critical approach to science and philosophy* (pp. 189–211). Free Press.

Aristotle. (1961 [4th century BCE]). *Metaphysics*, Books I–IX, Books X–XIV, Tredennick, H. (trans.), Loeb Classical Library. Cambridge, MA: Harvard University Press.

Boudry, M. (2011). *Here be dragons: Exploring the hinterland of science*. Ghent, Belgium: Ghent University. Faculty of Arts and Philosophy. https://biblio.ugent.be/publication/1191286

Boudry, M. (2013). Loki's wager and Laudan's error. On genuine and territorial demarcation. In M. Pigliucci & M. Boudry (Eds.), *Philosophy of pseudoscience: Reconsidering the demarcation problem* (pp. 79–98). The University of Chicago Press.

Boudry, M. (2017). Plus ultra. Why science does not have limits. In M. Pigliucci & M. Boudry (Eds.), *Science unlimited? The challenges of scientism* (pp. 31–52). University of Chicago Press.

Boudry, M., & Leuridan, B. (2011). Where the design argument goes wrong: Auxiliary assumptions and unification. *Philosophy of Science, 78*(4), 558–578.

Bunge, M. A. (1987). Why parapsychology cannot become a science. *Behavioral and Brain Sciences, 10*(4), 576–643.

Bunge, M. A. (2013). *Medical philosophy: Conceptual issues in medicine*. World Scientific.

Clarke, S. (2009). Naturalism, science and the supernatural. *Sophia, 48*, 127–142.

Coyne, J. A. (2009). *Why evolution is true*. Viking.

Dawkins, R. (2009). *The greatest show on Earth: The evidence for evolution*. Bantam.

Dembski, W. (2000). Naturalism and design. In W. L. Craig & J. P. Moreland (Eds.), *Naturalism: A critical analysis* (pp. 253–279). Routledge.

Dunkelberg, P. (2003). Irreducible complexity demystified. *Talk Reason*. https://www.talkdesign.org/faqs/icdmyst/ICDmyst.html

Fales, E. (2013). Is a science of the supernatural possible? In M. Pigliucci & M. Boudry (Eds.), *Philosophy of pseudoscience: Reconsidering the demarcation problem* (pp. 247–262). The University of Chicago Press.

Fishman, Y. I. (2009). Can science test supernatural worldviews? *Science & Education, 18*, 813–837.

Fishman, Y. I., & Boudry, M. (2013). Does science presuppose naturalism (or, indeed, anything at all)? *Science & Education, 22*(5), 921–949.

Gish, D. T. (1995). *Evolution: The fossils still say no!* Institute for Creation Research.

Hansson, S. O. (2013). Defining pseudoscience and science. In M. Pigliucci & M. Boudry (Eds.), *Philosophy of pseudoscience: Reconsidering the demarcation problem* (pp. 61–78). The University of Chicago Press.

Kitcher, P. (1981). Explanatory unification. *Philosophy of Science, 48*(4), 507–531.

Laudan, L. (1982). Commentary: Science at the bar—causes for concern. *Science, Technology, & Human Values, 7*(41), 16–19.

Mc Mullin, E. (1993). Evolution and special creation. *Zygon, 28*(3), 299–335.

National Academy of Sciences. (1999). *Science and creationism: A view from the National Academy of Sciences* (2nd ed.). The National Academies Press.

Paley, W. (1809). *Natural theology, or, evidences of the existence and attributes of the deity, collected from the appearances of nature* (12th ed.). Rivington.

Plantinga, A. (1991). When faith and reason clash: Evolution and the Bible. *Christian Scholar's Review, 21*, 8–32.

Pennock, R. T. (1999). *Tower of babel: The evidence against the new creationism*. MIT Press.

Quine, W. V. O. (1961). *From a logical point of view* (2nd ed.). Harper & Row.

Reisch, G. A. (1998). Pluralism, logical empiricism, and the problem of pseudoscience. *Philosophy of Science, 65*(2), 333–348.

Sober, E. (2008). *Evidence and evolution: The logic behind the science*. Cambridge University Press.

Worrall, J. (2004). Science discredits religion. In M. L. Peterson & R. J. VanArragon (Eds.), *Contemporary debates in philosophy of religion* (pp. 59–71). Blackwell.

Chapter 6
The Properties of Homeopathy

Abstract This chapter provides a brief definition and a description of/introduction to homeopathy. It then meta-analytically shows that the demarcation criteria and the scoring method proposed in Chap. 2 demarcate homeopathy as non-scientific, in agreement with the current philosophical and scientific consensus.

Keywords Homeopathy · Scientific demarcation · Non-science · Pseudoscience · Scientificity

6.1 Introduction

This chapter shows that the demarcation criteria demarcate homeopathy as non-scientific, in agreement with the current philosophical and scientific consensus. This section defines homeopathy. Section 6.2 lists properties of scientificity that homeopathy has. Section 6.3 lists properties of scientificity that homeopathy lacks. As the conclusion of this chapter, Sect. 6.4 argues that the demarcation criteria listed in Sect. 2.2 demarcate homeopathy as non-scientific (that homeopathy does not have enough of the properties of scientificity, and therefore should be deemed non-scientific), and that the fact that homeopathy is commonly deemed non-scientific is explained by the fact that it does not have enough of the properties of science listed in Chap. 2.

I take the properties of science listed in Sect. 2.2, and show that homeopathy lacks many of them. This provides: a) a rational reconstruction (and therefore an explanation) of why homeopathy is non-scientific, b) further evidence of the extensional adequacy of this work, c) further evidence of this work's explanatory power, and d) further evidence of the fact that the proposed set of criteria demarcates science appropriately.

Homeopathy (or homoeopathy), first proposed in 1796 by Samuel Hahnemann, is a system of alleged medicine that purports to heal patients by treating them with dilutions (or, for insoluble substances, triturations) of allegedly medicinal substances. Among the general population, homeopathy is frequently confused with

D. Fernandez-Beanato, *A Working Scientific Demarcation*, Synthese Library 519, https://doi.org/10.1007/978-3-031-99005-2_6

any type of herbal or "natural" medicine, but this constitutes a fundamental misunderstanding of the nature of homeopathy (cf. Shelton, 2004, p. 62), which is much more specific than that. Theoretically, modern homeopathy is based on two main tenets, namely, i) that these dilutions or triturations, together with their succussion (rapping, shaking, or stirring), increase the potency of the preparations; and ii) that some substances that cause symptoms of a disease in healthy persons can cure similar symptoms in sick patients (the Principle of Similars, *similia similibus curentur*, "like is cured by like", which gives its name to homeopathy). A third previously defining belief, put forward by Hahnemann, was that chronic diseases are caused by miasms, but this latter point has been discarded by some within homeopathy (Shelton, 2004, pp. 148–9).

Homeopathy's dilutions sometimes go beyond Avogadro's number; that is, they sometimes result in diluent excipients (usually water, ethanol, or sugar) in which it is statistically unlikely that there is a single molecule of the original substance in any given homeopathic dose, and therefore any allegedly active ingredient with which to treat the patient, even if the original substance had actual "active" medical components to start with. In modern times, this is sometimes justified within the homeopathic field by invoking "the [unsubstantiated] notion that ultra-dilutions can maintain an imprint of substances previously dissolved in them" (National Health Service, 2018). As Smith indicated,

> the idea that water (or any other diluents, such as alcohol) has a 'memory' is unsubstantiated by any known laws or mechanisms of chemistry or physics. More fundamentally, if it were true that water can 'remember' (perhaps by a yet-to-be-discovered revolutionary new scientific mechanism), then every sub- stance ever encountered by a molecule of water – such as food molecules, toxins, heavy metals – should be expected to imprint a 'memory' that would exert powerful and unpredictable effects on the body. The fact that physiological chaos does not ensure whenever we drink a glass of water provides a simple but powerful refutation of the 'memory' concept in homeopathy. (Smith, 2011, p. 3)

Homeopathy is used to treat very diverse ailments, both physical and psychological, including arthritis, asthma, depression, headaches, insomnia (Baran et al., 2014, p. 41), eating disorders, the attention deficit hyperactivity disorder (ADHD), compulsions, and many others, with some homeopaths even claiming to treat cancer.

After centuries of research, there is "no good-quality evidence that homeopathy is effective as a treatment for any health condition" (Ernst, 2002; Shang et al., 2005; quote: National Health Service, 2018), and therefore the continued belief in, or promotion of, homeopathy as a science or as scientific is a pseudoscientific attitude. Homeopathy is another CF commonly considered to be non-scientific and pseudoscientific by both philosophers and scientists (e.g., Tuomela, 1987, pp. 93–4, 99; Dawkins, 2003; Smith, 2012; Cleland & Brindell, 2013, pp. 183–4; Hansson, 2013, pp. 69–70; Koertge, 2013, p. 178; Ladyman, 2013, pp. 49, 53; Mahner, 2013, pp. 30, 35; Pigliucci & Boudry, 2013, p. 3; Baran et al., 2014, p. 41; Pigliucci, 2015, p. 572 ("one of the clearest examples of pseudoscience"); Mukerji & Ernst, 2022). In this way, rather than being a complementary system of medicine, homeopathy is a system of pseudo-medicine, and its alternativity to scientific medicine is the most immediate of its dangers (National Health Service, 2018).

Let us see how homeopathy stands regarding the possession of our scientificity properties.

6.2 The Properties of Homeopathy

We can concede that the best versions of homeopathy are falsifiable (at least in their claim that, ultimately, an above-chance or statistically significant percentage of the patients will get better) ("homoeopathy is eminently falsifiable", Mukerji & Ernst, 2022) (property 54).

6.3 The Properties that Homeopathy (Mostly) Does Not Have

Homeopathy is not empirically adequate ("the best clinical evidence has demonstrated no efficacy" (Sampson, 1995; Ernst, 2010; quote: Smith, 2011, p. 3)) (lack of property 183). Its results have not been consistently and successfully repeated/replicated in clearly high-quality clinical trials (Linde & Melchart, 1998; Shelton, 2004, pp. 11, 99, 197; Shang et al., 2005; Smith, 2011, p. 2; Mathie et al., 2014) (lack of property 57). The more rigorous the studies, the less positive the results are for homeopathy (Linde & Melchart, 1998; Shelton, 2004, pp. 11, 197; Shang et al., 2005, p. 730). "[T]he effects seen in placebo-controlled trials of homoeopathy are compatible with the placebo hypothesis" (quote: Shang et al., 2005, p. 730; Smith, 2011, p. 3; Grimes, 2012, p. 149). "Homeopathy is still looking for a plausible mechanism that explains how homeopathic remedies should work, particularly the many very dilute remedies with none of the original […] molecules remaining" (Shelton, 2004, pp. 10 (quote), 57, 69–70) (lack of property 179). It features a low degree of theoretical confirmation (lack of property 184) and lacks bulk scientific external consistency (Shelton, 2004, pp. 57, 63; Shang et al., 2005; Grimes, 2012): none of the above-mentioned three tenets of homeopathy are substantiated by, or compatible with, current science. "It is pharmacologically untenable to claim that the smaller the dose the bigger the effect" (Smith, 2012, p. 2), because "[e]mpirically, pharmacology has demonstrated that physiological responses are dose-dependent" (Smith, 2011, p. 3). And "there is simply no plausible or rational basis for supposing that substances that produce particular symptoms can cure the same symptoms" (Smith, 2011, p. 2), "we now know that […] [t]he only correct [curative] therapy for an illness is according to its very specific etiology and pathogenesis" (Smith, 2012, p. 2). "[F]or homeopathic dilutions to have mechanistic effects, it would appear necessary to reject virtually all that science has painstakingly elucidated over the last 200 years concerning the composition of matter itself" (Smith, 2012, p. 3)) (lack of property 190).

Homeopathy neither measures nor quantifies enough (Shelton, 2004, pp. 68–9, 176) (lack of property 92). It uses and cites unrepresentative, handpicked ("cherry-picked") data, evidence, examples, statements, studies, or outcomes (Smith, 2012, pp. 1–3; Mukerji & Ernst, 2022) (lack of property 97). It does not rely on clearly high-quality, well-done clinical trials (Shelton, 2004, pp. 73, 98) (lack of property 96). It over-relies on figures of authority, specifically Hahnemann and his writings, to settle disputes on matters of fact (Shelton, 2004, pp. 11, 69) (lack of property 108). Regarding the strictures of "contemporary research methods", homeopaths believe that "their presence or absence should not be used to disqualify a homeopathic drug proving strictly done according to the directions set forth by Hahnemann" (Homeopathic Pharmacopoeia of the United States, 2001). The community of homeopaths is not interested in properly testing their theory by putting its claims to meaningful, well-conducted tests (lack of property 206). Regarding clinical trials, it is unjustifiedly selective when collecting data and considering confirmations and disconfirmations (Smith, 2012, pp. 1–3) (lack of property 78).

Not all its procedurally valid studies give consistent results (lack of property 211), with some outlying clinical studies supporting homeopathy. The latter are explainable "by a range of factors including chance statistical effects, low subject numbers, poor study design, slipshod execution, and even fraud" (Smith, 2012, p. 2). The community of homeopaths does not self-correct (Smith, 2012) (lack of properties 104 and 217).

Homeopathy does not predict well. It "does not predict when, how, and to what extent each patient will get better. Patient response is said to be highly individual—that is, unpredictable. [...] Many homeopaths agree that the results of homeopathic treatment are not very predictable" (Shelton, 2004, p. 170) (lack of properties 45, 82, and 184).

> Given the enormous variety of remedy selection methods, a group of homeopaths is guaranteed to select many different remedies for the same patient, even homeopaths practicing classical homeopathy (Shelton, 2004, p. 294) (lack of properties 29 and 81).
>
> [H]omeopathy gives more credence to individual cases than to large clinical trials. Case reports are a major part of most homeopathy books, courses, journals, magazines, and conferences. Homeopaths tell skeptics that if the skeptics experience homeopathy they will believe in it (Shelton, 2004, pp. 187–96, quote: 187) (lack of properties 132–133).
>
> It is rare that homeopaths consider any explanation of their patients' progress other than the remedy (Shelton, 2004, p. 175) (lack of properties 74, 152, and 212).

Homeopathy uses unfounded immunisations, is unconnected to science, lacks evidence, reverses the burden of proof, and mischaracterises evidence (Mukerji & Ernst, 2022) (properties 12, 57, 100–101, and 155).

Regarding all the properties in Sect. 2.2 not mentioned in this chapter, it is unclear whether homeopathy mostly has them.

6.4 Chapter Conclusions

Homeopathy lacks at least 28 of the properties of scientificity featured in the demarcatory list, while not clearly having any specific one. This yields a score of −28 for homeopathy. This (possibly provisional) tally is strongly against the scientificity of homeopathy, and is the best explanation for the fact that homeopathy is commonly deemed non-scientific. This, once again, demarcates in a manner that is equivalent to that of the current international philosophical and scientific consensus regarding the extension of the term "science".

References

Baran, G. R., Kiana, M. F., & Samuel, S. P. (2014). Chapter 2: Science, pseudoscience, and not science: How do they differ? In G. R. Baran, M. F. Kiana, & S. P. Samuel (Eds.), *Healthcare and biomedical technology in the 21st century* (pp. 19–57). Springer.

Cleland, C. E., & Brindell, S. (2013). Science and the messy, uncontrollable world of nature. In M. Pigliucci & M. Boudry (Eds.), *Philosophy of pseudoscience: Reconsidering the demarcation problem* (pp. 183–202). The University of Chicago Press.

Dawkins, R. (2003). *A devil's chaplain.* Weidenfeld & Nicolson.

Ernst, E. (2002). A systematic review of systematic reviews of homeopathy. *British Journal of Clinical Pharmacology, 54*(6), 577–582.

Ernst, E. (2010). Homeopathy: What does the 'best' evidence tell us? *Medical Journal of Australia, 192*, 458–460.

Grimes, D. R. (2012). Proposed mechanisms for homeopathy are physically impossible. *Focus on Alternative and Complementary Therapies, 17*(3), 149–155.

Hansson, S. O. (2013). Defining pseudoscience and science. In M. Pigliucci & M. Boudry (Eds.), *Philosophy of pseudoscience: Reconsidering the demarcation problem* (pp. 61–78). The University of Chicago Press.

Homeopathic Pharmacopoeia of the United States. (2001). *Abstracts 2001 (HPUS).* Homeopathic Pharmacopoeia Convention of the United States.

Koertge, N. (2013). Belief buddies versus critical communities: The social organization of pseudoscience. In M. Pigliucci & M. Boudry (Eds.), *Philosophy of pseudoscience: Reconsidering the demarcation problem* (pp. 165–180). The University of Chicago Press.

Ladyman, J. (2013). Toward a demarcation of science from pseudoscience. In M. Pigliucci & M. Boudry (Eds.), *Philosophy of pseudoscience: Reconsidering the demarcation problem* (pp. 45–60). The University of Chicago Press.

Linde, K., & Melchart, D. (1998). Randomized controlled trials of individualized homeopathy: A state-of-the-art review. *Journal of Alternative and Complementary Medicine, 4*(4), 371–388.

Mahner, M. (2013). Science and pseudoscience. How to demarcate after the (alleged) demise of the demarcation problem. In M. Pigliucci & M. Boudry (Eds.), *Philosophy of pseudoscience: Reconsidering the demarcation problem* (pp. 29–44). The University of Chicago Press.

Mathie, R. T., Lloyd, S. M., Legg, L. A., Clausen, J., Moss, S., Davidson, J. R. T., & Ford, I. (2014). Randomised placebo-controlled trials of individualised homeopathic treatment: Systematic review and meta-analysis. *Systematic Reviews, 3*, 142.

Mukerji, N., & Ernst, E. (2022). Why homoeopathy is pseudoscience. *Synthese, 200*, 394.

National Health Service. (2018). Health A-Z—homeopathy. https://www.nhs.uk/conditions/homeopathy/

Pigliucci, M., & Boudry, M. (2013). Introduction. In M. Pigliucci & M. Boudry (Eds.), *Philosophy of pseudoscience: Reconsidering the demarcation problem* (pp. 1–6). The University of Chicago Press.
Pigliucci, M. (2015). Scientism and pseudoscience: A philosophical commentary. *Journal of Bioethical Inquiry, 12*, 569–575.
Sampson, W. (1995). Homeopathy does not work. *Alternative Therapies in Health and Medicine, 1*(4), 48–52.
Shang, A., Huwiler-Müntener, K., Nartey, L., Jüni, P., Dörig, S., Sterne, J. A. C., Pewsner, D., & Egger, M. (2005). Are the clinical effects of homoeopathy placebo effects? Comparative study of placebo-controlled trials of homoeopathy and allopathy. *The Lancet, 366*(9487), 726–732.
Shelton, J. W. (2004). *Homeopathy: How it really works*. Prometheus Books.
Smith, K. (2011). Against homeopathy—a utilitarian perspective. *Bioethics, 26*(8), 398–409.
Smith, K. (2012). Homeopathy is unscientific and unethical. *Bioethics, 26*(9), 508–512.
Tuomela, R. (1987). Chapter 4: Science, protoscience, and pseudoscience. In J. C. Pitt & M. Pera (Eds.), *Rational changes in science: Essays on scientific reasoning. Boston studies in the philosophy of science* (Vol. 98) (pp. 83–101). Springer.

Chapter 7
The Properties of Feng Shui

Abstract This chapter provides a brief definition and a description of/introduction to feng shui. It then meta-analytically shows that the demarcation criteria and the scoring method proposed in Chap. 2 demarcate feng shui as non-scientific, in agreement with the current philosophical and scientific consensus.

Keywords Feng Shui · Geomancy · Topomancy · Scientific demarcation · Non-science · Pseudoscience · Scientificity

7.1 Introduction

This chapter shows that the demarcation criteria demarcate feng shui as non-scientific, in agreement with the current philosophical and scientific consensus. This section defines feng shui. Section 7.2 lists properties of science that feng shui has. Section 7.3 lists properties of science that feng shui lacks. As the conclusion of this chapter, Sect. 7.4 argues that the demarcation criteria listed in Sect. 2.2 demarcate feng shui as non-scientific (that feng shui does not have enough of the properties of scientificity, and therefore should be deemed non-scientific), and that the fact that feng shui is commonly deemed non-scientific is explained by the fact that it does not have enough of the properties of science listed in Chap. 2.

I take the properties of science listed in Sect. 2.2, and show that feng shui lacks many of them. This provides: a) a rational reconstruction (and therefore an explanation) of why feng shui is non-scientific, b) further evidence of the extensional adequacy of this work, c) further evidence of this work's explanatory power, and d) further evidence of the fact that the proposed set of criteria demarcates science appropriately.

Feng shui is a CF that is commonly considered to be non-scientific and pseudoscientific by both philosophers and scientists (Sullivan, 1990, p. 264; Tseng et al., 2013, p. 6; Matthews, 2018, pp. 31, 33, 2019a, 2019b; Hong et al., 2020).

This chapter was previously included in Fernandez-Beanato (2021).

D. Fernandez-Beanato, *A Working Scientific Demarcation*, Synthese Library 519, https://doi.org/10.1007/978-3-031-99005-2_7

Feng shui (traditional Chinese characters: 風水, simplified: 风水, Hanyu Pinyin: fēngshuǐ), also known as siting and as Chinese geomancy or topomancy, is a traditional practice purported to be able to "harmonise" individuals with their surrounding environments by taking into consideration certain energy forces or fields, therefore promoting the individuals' well-being. It covers the locations and orientations of cities, villages, farms, buildings, homes, and, traditionally, tombs and graves, their characteristics, structure, architecture, layout, configurations, arrangements, proportions, contents, decoration, the positions and smooth functioning of their internal moving parts, lighting, gardening, and aromatisation.

The main one of these energies or forces, called chi or ch'i (traditional: 氣, simplified: 气, Hanyu Pinyin: qì), is identified as a primal, universal, and flowing life-force or vital energy as yet undiscovered by modern science, or, alternatively, as an entity that already has a standing in modern science, for example, "electromagnetic energy" (e.g., Brown, 2005, p. 24).[1] To cover both of these ontological versions, this work assesses feng shui independently of particular ontologies of the energies or forces in question. This course of action is fruitful since the belief in the alleged results of feng shui practices, which is the common denominator of all versions, is sufficient to assess feng shui as a whole.

Here, "feng shui" is not to be understood in a wider sense, i.e., as including any approach directed to achieve some type of "human–environment harmony". Such a definition of "feng shui", besides being non-standard, would be too wide, as it would include scientific approaches guided by considerations regarding ecology, conservation, sustainability, psychology, habitat planning, etc., and featuring no reference to the pseudoscientific beliefs discussed here (see, e.g., Lau et al., 2005). "Feng shui" is not just another name for the study of "environmental protection and management, energy conservation, renewable energy, sustainability and ecology" (cf. Mak & So, 2015, p. xxvii).

The wide definition would also include practices guided by mere common sense; but, for example, mere common-sense architecture is not feng shui. The availability of water, one of the most important principles of feng shui, is obviously essential for human life. "Contentment goes along with living by a lake with nice views" (Matthews, 2019b, p. 264), and a restaurant with a pleasing sea view will attract more customers, because these are "psychologically attractive setting[s]" (Emmons, 1992, p. 48). This is probably why some principles of feng shui "can obviously be explained by theories of modern environmental science" (Mak & So, 2015, p. xxviii); why, if Han (1995) is correct, major landscape criteria in environmental psychology are similar to criteria in the Form (or Shapes) School of feng shui; and why Lynch (2003) and Mak and Ng (2005) found that architects and designers design the way feng shui experts would advise them to design (although some of the professionals consulted might have been believers in feng shui) (cf. March, 1968, pp. 254–5). The main findings in So and Lu (regarding airflow) (2001) and those in

[1] Feng shui is just one of many disciplines or practices allegedly using chi. Other examples are qìgōng, reiki, so-called traditional Chinese medicine, martial arts, and different Taoist practices.

Low et al. (2012) could be other examples of calling unrelated successes "feng shui". Contentment also goes along with places, features, arrangements, or configurations favoured by one's culture or beliefs.

The claim that feng shui is not scientific is a specific, technical one. It does not imply that there are no truths in feng shui, or that it does not make any good recommendations. Because of common sense and of socially accumulated environmental experience, feng shui does make recommendations for the siting of villages that are advantageous with regard to self-protection and the use of natural resources such as water and land (see Anderson & Anderson, 1973, pp. 45–50; Yoon, 2009).

An intuition of interconnectedness and interdependence between humans and their surrounding environments is coherent, rational, reasonable, and correct according to current physics, chemistry, biology, ecology, economics, and psychology. So, in some sense, Emmons (1992, p. 40) is wrong to say that "the notion of living in harmony with environmental elements appears to be entirely a kind of magic". Said intuition could even arguably be considered protoscientific in a premodern context (for example, as cosmological theory, as in Liu, 2015, p. 33). This, and the above-mentioned empirical successes, are reasons why the claim that feng shui is not scientific does not imply that feng shui lacks cultural value, or that it cannot be a good alternative in some contexts. But the temporal dimension, the social or cultural context, and the availability of epistemically better or worse alternatives should play their parts in assessments of pseudoscientificity. A protoscience is protoscientific only as long as it is the CF that contextually (in a given period and society) has the most properties of science on the topics and aspects of reality that it covers. In this way, previous developments stop being scientific as soon as a (demonstrably) sufficiently epistemically-progressive integral separated approach develops. Feng shui has become a pseudoscientific CF in modern times because of its unjustified continuity in the face of much more scientific available alternatives. Therefore, the continued belief in, or promotion of, feng shui (with all of its theoretical posits and propositions regarding the empirical) as a science or as scientific is a pseudoscientific attitude.

Let us see how feng shui stands regarding the possession of our scientificity properties.

7.2 The Properties of feng shui

The best versions of feng shui have the following properties:

Feng shui is a naturalistic holistic conception that purports to refer primarily to legal workings of the system of nature or of the cosmos ("There is one truth in Feng-shui. [...] It is the recognition of the uniformity and universality of the operation of natural laws." (Eitel, 1873/1987, p. 82)). This, and its intuition of interconnectedness and interdependence between humans and their surrounding environments, appear to be why some scholars have said that feng shui is (or was) rudimentary natural science (Eitel 1873 [1987]), or that it has a scientific aspect (Yoon, 2009). Therefore, it possesses properties 116 and 181–182.

But feng shui can also use unfounded immunisations to avoid or evade disconfirmations/falsifications. For an example, see its lack of property 97, below in Sect. 7.3.

7.3 The Properties that feng shui (Mostly) Does Not Have

Feng shui lacks predictive success (lack of property 82) (Matthews, 2019b, p. 297) beyond the advantageous recommendations mentioned in Sect. 3.1 (assessments on the predictive success of a CF should be done overall and with attention to the possibility of the existence of epistemically better alternative explanations, because almost every belief system predicts successfully in some respect some of the time, e.g., believers in the Bermuda Triangle successfully predict that ships and aeroplanes will be lost in that area). There are no studies published in reputable peer-reviewed scientific journals showing feng shui's differential[2] empirical adequacy or success in any sufficiently relevant and specific predictions. As e.g. in the cases of parapsychology and genethliac astrology, apparently confirmatory/corroboratory experimental results or successful predictions are reported from time to time. But only a consistent record of success (without an epistemically better explanation) by independent investigators can be accepted for deeming that a CF possesses predictive success, or that a claim has been experimentally confirmed/corroborated conclusively. This demand for consistency is because differences in interpretation, flaws in observational or experimental design, mistakes due to cognitive errors and biases, fraud, etc., can occur. Scientific experiments on feng shui are not usually carried out (Eitel, 1873/1987, p. 69; Matthews, 2019b, pp. 288, 293). But when they are, their outcomes do not support feng shui. "[T]he testing that has been done confirms what is obvious: […] the bulk of […] confirmations are simply […] affirming the consequent" (Matthews, 2019a, p. 4, 2019b, p. 267). Affirmations of the consequent (i.e., arguments of the form P → Q, Q, therefore P) are routine in science, but they should not be used when epistemically better explanatory alternatives are available. Feng shui "claims have never been confirmed in any reputable laboratory test" (Matthews, 2019a, p. 4). Also, in feng shui, "[u]nrepeatable experiments are too frequently adduced" (Matthews, 2018, p. 31) ("unrepeatable" in the sense that they are in-principle repeatable but cannot be empirically replicated) (Matthews, 2018, pp. 26–7).

Feng shui is not empirically adequate (lack of property 183) beyond the advantageous recommendations mentioned in Sect. 3.1 (assessments on the empirical adequacy of a CF should be done overall, because almost every belief system is empirically adequate in some respect some of the time, e.g., genethliac astrology

[2] "Differential" in the sense of being better at predicting than well-established and bulk-externally-consistent scientific theories.

predicts successfully sometimes, and homeopaths cause remissions via the placebo effect). There are no differentially confirming/corroborating accounts of feng shui published in reputable peer-reviewed scientific journals (lack of property 5). As in the case of genethliac astrology, the confirmation/corroboration of feng shui would require a sample of cases that is sufficiently large and representative enough to show that, in general, the number of cases that confirm the predictions of feng shui tends to be higher than the number that is expected to happen by mere chance or for other well-known reasons, that is, a number higher than the number that would occur if the feng shui theory were false. Ideally, this would take the form of large-scale statistical studies.

Therefore, feng shui lacks theoretical confirmation (success of tests or of predictions, lack of property 184) (cf. Eitel, 1873/1987, p. 69).

Feng shui sometimes resorts to unfounded immunising stratagems (lack of properties 7 and 12–13) or "'get out of jail' cards" (Matthews, 2019b, p. 294). These are stratagems that feng shui theoreticians and practitioners employ to avoid or evade disconfirmation/falsification (in this way, the impression that most results are confirmatory/corroboratory of feng shui can be maintained; in more extreme cases, these recourses can make feng shui immune to testing). For example, feng shui claims that taking certain feng shui measures indicated for prosperity can bring money to interested parties, but not necessarily, because prosperity has to be understood in a wider sense, as welfare, such as having sufficient food and heat. But then again, prosperity should also be understood in an even wider sense, as sustainability, that is, the ability to maintain the conditions that can lead to welfare, which in turn can lead to having money. Sustainability does not necessarily entail welfare, in the same sense that welfare does not necessarily entail monetary wealth. And, in any case, it is up to the individual to take the relevant opportunities to earn money. So, if the individual does not get much money, it can be that she did not make a good enough use of the available opportunities. And even if it could be proved that she did (which is exceedingly difficult), the fact that she did not earn enough money is still not a disconfirmation/falsification of feng shui, because it could still be the case that her karma is awful.

The karmic stratagem has the power to render feng shui unfalsifiable (empirically invulnerable), because karma is an inscrutable factor. In any case, any defender of feng shui who is concerned with epistemic intersubjectivity or with epistemically justified acceptance of feng shui claims must contend that feng shui predictions are effectively fulfilled in a number of cases. This number has to be higher than expected by mere chance if there is going to be a non-mystical way of knowing that feng shui works. But this is inscrutable at present because not enough large-scale (observational or experimental) scientific studies of feng shui's alleged effects have been conducted, if any.

The question of whether feng shui's use of these subterfuges renders feng shui altogether unfalsifiable is a complex one. Narrowly-construed and definite feng shui claims, without reference to karma or to other immunising subterfuges, are testable and falsifiable, which is why they have been falsified. This is why this assessment does not count the property of falsifiability for nor against feng shui.

Feng shui is isolated and non-externally-systematic in that it is not, in Bunge's words, "a component of the system of the sciences" (Matthews, 2019b, p. 280; quote: Bunge, 1987, p. 576). Feng shui is not integrated with any of the most successful uncontroversially scientific CFs, it is neither tied up nor intertwined with physics, chemistry, or biology. Neither the specific concepts (t-theoretical concepts) of the different theoretical feng shui views on why feng shui would work, nor those theories themselves, are integrated into uncontroversial science. This makes feng shui lack properties 57, 105, 189, 194, and 208.

Feng shui is bulk externally inconsistent (lack of property 190) (Matthews, 2018, p. 33, 2019b, pp. 293–4) in many ways. One example is its use of scientific terms without regard for their scientific meanings (cf. Matthews, 2019b, p. 235). Empirically, feng shui is inconsistent e.g. with basic ceteris paribus assumptions in biology, psychology, and economy which indicate that no other relevant factors are at work influencing degrees of health or prosperity.

Because of the lack of predictive success, empirical adequacy, and experimentation (testing) covered above in this section, it can be concluded that feng shui does not abide by evidence (lack of property 96).

Feng shui features "cherry-picking" (lack of properties 97 and 100) (Matthews, 2018, p. 31). Because the CF of feng shui is forced to present an appearance of empirical adequacy in the face of the above-mentioned predictive failures if it wants to be deemed legitimate, feng shui theoreticians and practitioners pick and choose results (see e.g. the discussion about unfounded immunising stratagems above in this section).

As covered above in this section, feng shui does not perform enough scientific observation or experimentation (this in turn indicates that feng shui theoreticians, practitioners, and believers have no desire for scientific vindication (Matthews, 2018, p. 31, 2019a, p. 5)). Feng shui is also not (self-)critical (lack of property 118) (Matthews, 2018, p. 31, 2019b, pp. 257–8). Self-criticalness would be best implemented in this case via a more sceptical attitude accompanied by large-scale scientific self-check, including objective observation and experimentation with discrimination of variables and consideration of alternative scientific explanations, which are already available and bulk externally consistent. As covered above in this section, feng shui falls short on all of these fronts. Also, as covered above in this section, it is deemed acceptable in the CF of feng shui to explain away disconfirmations/falsifications by recourse to unfounded immunising stratagems. All this means that "disconfirmation is neither sought nor recognized" (Matthews, 2018, p. 31) in feng shui (lack of properties 7, 99, 120, and 206–207).

Feng shui does not primarily use the general scientific method as its main strategy for dealing with its cognitive problematics (lack of property 104) (cf. Matthews, 2019b, p. 293). From the analysis above in this section, it can be seen that this is especially notable in feng shui's lack of systematic observation and experimentation (in Bunge's terms, feng shui falls short particularly on the methodic steps of checking its solution to cognitive problems and evaluating it in the light of both test and background knowledge).

Feng shui makes "no effort to disentangle variables and study their contributions" (lack of property 112) (Matthews, 2019b, p. 293). Such disentanglement would be necessary if feng shui conducted scientific (analytic) observation or experimentation. Many authors, including Needham, have noted that feng shui was born out of a holistic worldview; this should not be, and is not, an obstacle to the possibility of testing feng shui scientifically (which is better done by conceptually disentangling variables).

Feng shui remains stagnant in the face of contradictory evidence. This is in part because the feng shui community is willing to use unfounded immunising stratagems (see above in this section) to avoid accepting disconfirmations/falsifications. Feng shui does not progress. Its content of knowledge does not grow (lack of properties 14, 83–84, and 217). "[D]espite 3000–4000 years of adherence and cultivation, there is no cognitive growth" (Matthews, 2019b, pp. 273–4, quote: 293–4).

As indicated above in this section for the features of predictive success, empirical adequacy, and experimentation, feng shui claims are not "reported in any mainstream science journal", and "there are no contributions to established, [scientifically] peer-reviewed, scientific research journals" (Matthews, 2019a, p. 4, 2019b, p. 293) that have come from the feng shui field (lack of property 128).

Given the above-mentioned lack of sufficient scientific observation or experimentation, and the existence of failures in empirical adequacy and prediction, no scientific bases exist for believing that most empirical results are confirmatory/corroboratory of feng shui, or for choosing mostly confirmatory results as the bases for believing in feng shui or in its reliability. This is because in such situations, believers can be prey to cognitive errors and biases, including confirmation bias (which is one of the reasons why scientific validation is necessary). The existence of said failures indicates that continued belief in feng shui requires that believers be prey to some of those errors and biases (lack of property 147) (Matthews, 2018, p. 31).

In its disregard for scientific observation, experimentation, and in its "cherry-picking", feng shui ignores criteria for obtaining evidence validly (lack of property 154).

By its use of unfounded immunising stratagems, feng shui appeals to unfounded non-mysteries, in that it deliberately creates mystery where none exists by omitting crucial information on disconfirmations (lack of properties 25–26) (Eitel, 1873/1987, p. 1; Matthews, 2019b, pp. 265, 288).

Disregarding scientific observation and experimentation in empirical areas, as well as scientific validation and (self-)criticalness, is not conducive to reliable or truthlike beliefs about reality. Therefore, feng shui beliefs do not conform with reasons for believing, which amounts to irrationality (lack of property 65).

Regarding all the properties in Sect. 2.2 not mentioned in this chapter, I have not encountered in the academic literature reviewed an analysis of feng shui considering the possession or lack of those properties, but it appears to be clear that feng shui lacks most of them.

7.4 Chapter Conclusions

Feng shui lacks at least 33 of the properties of science featured in the demarcatory list, while having at least 3 of the specific ones. This yields a score of −30 for feng shui. This (possibly provisional) tally is strongly against the scientificity of feng shui, and is the best explanation for the fact that feng shui is commonly deemed non-scientific. This is yet one more case in which my demarcatory procedure demarcates in a manner that is equivalent to that of the current international philosophical and scientific consensus regarding the extension of the term "science".

References

Anderson, E. N., & Anderson, M. (1973). Changing patterns of land use in rural Hong Kong. In E. N. Anderson & M. Anderson (Eds.), *Mountains and water: Essays on the cultural ecology of south coastal China*. Taipei.

Brown, S. (2005). *The Feng Shui bible: The definitive guide to practising Feng Shui*. Octopus Publishing Group.

Bunge, M. A. (1987). Why parapsychology cannot become a science. *Behavioral and Brain Sciences, 10*(4), 576–643.

Eitel, E. J. (1873/1987). *Feng Shui: The rudiments of natural science in China*. Lane Crawford (Graham Brash, Singapore).

Emmons, C. F. (1992). Hong Kong's Feng Shui: Popular magic in a modern urban setting. *The Journal of Popular Culture, 26*(1), 39–50.

Fernandez-Beanato, D. (2021). Feng Shui and the demarcation project. *Science & Education, 30*(6), 1333–1351.

Han, K. T. (1995). *Basic theory of landscape Feng Shui*. Lamper Enterprises. In Chinese.

Hong, J.-C., et al. (2020). Undergraduate Science Students' Scientist–Practitioner Gap: the Role of Epistemic Curiosity and Cognitive Flexibility. *International Journal of Science and Mathematics Education.*

Lau, S. S., et al. (2005). Sustainable design in its simplest form: Lessons from the living villages of Fujian rammed earth houses. *Structural Survey, 23*(5), 371–385.

Liu, J. (2015). In defense of Chinese Qi-naturalism. In C. Li, F. Perkins, & A. K. L. Chan (Eds.), *Chinese metaphysics and its problems* (pp. 33–53). Cambridge University Press.

Low, S. P., Den, X., & Quek, L. T. (2012). Assimilating Total building performance mandates with Chinese geomancy principles and scenarios. *Facilities, 30*(13/14), 558–589.

Lynch, E. (2003). Feng Shui as a site design tool: Assessing conditions of human comfort in urban places. Unpublished M. L. A. dissertation. University of Arizona, USA.

Mak, M. Y., & Ng, S. T. (2005). The art and science of Feng Shui—a study on architects' perception. *Building and Environment, 40*(3), 427–434.

Mak, M. Y., & So, A. T. (2015). *Scientific Feng Shui for the built environment: Theories and applications*. City University of Hong Kong Press.

March, A. L. (1968). An appreciation of Chinese geomancy. *Journal of Asian Studies, 27*(2), 253–267.

Matthews, M. R. (2018). *History, philosophy and science teaching: New perspectives*. Springer.

Matthews, M. R. (2019a). *Feng Shui: Philosophical appraisal and educational opportunity*. HPS&ST Newsletter.

Matthews, M. R. (2019b). *Feng Shui: Teaching about science and pseudoscience*. Springer.

So, A. T. P., & Lu, J. W. Z. (2001). Natural ventilation design by computational fluid dynamics—a Feng Shui approach. *Architectural Science Review, 44*(1), 61–69.

Sullivan, M. (1990). Chinese art and its impact on the West. In P. Ropp (Ed.), *Heritage of China: Contemporary perspectives on Chinese civilization* (pp. 263–293). University of California Press.

Tseng, Y.-C., et al. (2013). The relationship between exposure to pseudoscientific television programmes and pseudoscientific beliefs among Taiwanese university students. *International Journal of Science Education, Part B, 4*(2), 107–122.

Yoon, H. (2009). Tracing rational aspects of Feng Shui (geomancy). In M. Y. Mak & A. T. So (Eds.), *Scientific Feng Shui for the built environment: Fundamentals and case studies*. City University of Hong Kong Press.

Chapter 8
The Properties of Common Sense

Abstract This chapter provides a brief definition and a description of/introduction to common sense. It then meta-analytically shows that the demarcation criteria and the scoring method proposed in Chap. 2 demarcate common sense as non-scientific, in agreement with the current philosophical and scientific consensus.

Keywords Common sense · Scientific demarcation · Science · Non-science · Psychology · AI · Scientificity

8.1 Introduction

We use common sense in our everyday lives to aim at gaining propositional knowledge. It is in this capacity that common sense is studied in philosophy and in the CF of artificial intelligence. Therefore, according to our definition in Note 11 of Chap. 1, common sense is a CF. As such, it can be evaluated with respect to the criteria of scientificity. As we do in the case of science, this chapter ignores all epistemically-irrelevant aspects and properties of common sense. The use of common sense as a source of knowledge is routinely successful because modern common sense has many of the epistemically-good properties listed in Sect. 2.2.1, and therefore modern common sense is an epistemically-good CF. Still, this chapter shows that the demarcation criteria demarcate common sense as non-scientific, in agreement with the current philosophical and scientific consensus.

The concern of this chapter is what properties of science common sense has and what properties of science it lacks. I take the properties of science listed in Sect. 2.2, and show that common sense possesses many of them. This provides: a) a rational reconstruction (and therefore an explanation) of *why* common sense is non-scientific, b) further evidence of the extensional adequacy of this work, c) further evidence of this work's explanatory power, and d) further evidence of the fact that the proposed set of criteria demarcates science appropriately.

This chapter also shows that epistemically virtuous common sense tends to possess many of the properties that are solely negatively demarcatory of science (i.e.,

D. Fernandez-Beanato, *A Working Scientific Demarcation*, Synthese Library 519, https://doi.org/10.1007/978-3-031-99005-2_8

the properties that are non-specific to science, listed in Sect. 2.2.1), and to lack the properties that are both negatively and positively demarcatory of science (i.e., the properties that are specific to science). In doing so, this chapter provides evidence for the appropriateness of this work's distinction between those two kinds of properties. Lacking any of the properties listed in Sect. 2.2.1 counts against something being epistemically-good common sense in the same sense that it counts against it being science or scientific, because it counts against it qua epistemically-good CF, which both epistemically-good common sense and science are.

The English expression "common sense" is polysemic. Sections 8.2–8.5 conduct a brief exploration of how the expression has been, and is, used in everyday language, philosophy, psychology, and artificial intelligence studies. This serves to clarify what this chapter means (and does not mean) by "common sense" when evaluating common sense against the properties of scientificity. This adopted meaning is specified in Sect. 8.6. Sect. 8.7 lists properties of scientificity that common sense has. Section 8.8 lists properties of scientificity that common sense lacks. Section 8.9 argues that because common sense does not have enough of the specific properties of scientificity, it should be deemed non-scientific.

8.2 The Expression "common sense" in Everyday Language

This section considers the uses of the expression "common sense" in everyday language. The rich polysemy of the expression "common sense" is apparently inherited from the polysemies of both the terms "sense" and "common". "Sense" can refer to a semantic entity, to a faculty, to sensory systems or mechanisms, to intelligibility, to plausibility, to adequacy, etc. "Common" can mean "universal", "widespread", "general", "generic", "ordinary", "mundane", "unsophisticated", "not enhanced", "unspecialised", etc. Judging from the main meanings of the expression "common sense" given in the Oxford English Dictionary, "common sense" can refer to at least four different types of things, namely:

(i) common-sense practices, methods, techniques, ways of thinking, of reasoning, of arguing, of judging, or of coming to conclusions, for forming beliefs, for gaining knowledge, or for arriving at conclusions (this is why common sense is a CF);
(ii) common-sense beliefs or propositions;
(iii) a level of judgement which is contextually less sophisticated than that of science; and
(iv) a faculty of judgement (cf. Bunge, 2003, p. 44) which is an overall ability to find out the facts about a situation and to reason.

In sum, the expression "common sense" is used to refer to: i) certain procedures, ii) the doxastic or epistemic results of those procedures, iii) the epistemic level at which those procedures occur, and iv) the epistemic faculty that makes those processes possible. All of these referents can coexist with each other. "Common sense" can also refer to mixtures of some or all of these things.

In all these four general senses (procedural, doxastic/propositional, as a level, and facultative), "common sense" involves "knowledge and rationality" that are "ordinary" (ibid.) for a given epistemic community, and what is common-sensical depends on physiology and on culture. Therefore, common sense can vary between cultures (and might vary between species). Some elements of common sense can change in time, or be simultaneously different in different epistemic communities, some of these changes and variations being merely due to common sense becoming loaded with scientific knowledge or scientific beliefs, or with beliefs of other kinds. Many widespread beliefs are products of science. And because these loadings can occur at different times in the development of science, common sense can feature mixtures of superseded science and current science (cf., e.g., Halloun & Hestenes, 1985, p. 1056). Scientifically superseded current common-sensical beliefs can serve as a contrast class for science because a comparison between their supersession in science and their continued acceptance in common sense would reveal the differences in the respective doxastic dynamics of common sense and of science. On the other hand, beliefs that are current both in common sense and in science cannot serve as a contrast class for science, because they are the same in both CFs. This is why this latter kind of beliefs is not taken into consideration when contrasting common sense with science in Sects. 8.7–8.8.[1]

When used in any of the above-mentioned four general senses, the expression "common sense" can be used in at least two different (but mutually compatible) more specific senses, i.e., a "cultural" sense, in which the referent (procedure, belief, level, or faculty) is deemed dependent on culture (independently of epistemically good backing, epistemic virtue, or epistemic success); and another sense, a "virtuous" sense in which the referent is considered to possess epistemic virtue or a tendency to achieve epistemic success. What follows in this section details these bifurcations for each of the four types of referents:

When referring to practices, methods, etc. (our referent (i)), the expression "common sense" can mean: i)a) that these practices, methods, etc. are widespread within a community, society, or culture (there may or may not be epistemically good reasons for adhering to these methods); or i)b) that these practices, methods, etc., besides being easily applicable by members of the relevant epistemic community, are generic, generally applicable (or even universally applicable), and generally epistemically good (effective, appropriate to the goal of knowing true propositions).[2]

---

[1] Science unavoidably rests, in part, on input obtained from common sense, for example in its research activities (e.g., scientists apply simple common-sensical procedures, such as looking, to read results of measurements off of some instruments). For our purposes here, we can be neutral on the question of whether the methodics of science is a refinement and perfecting of the epistemically reliable procedures of common sense (i.e., common sense in our sense i)b), or something qualitatively different. In either case, it is epistemically superior to common sense.

[2] It is in this sense that it is a cliché that "common sense is not common (widespread)". These general practices, methods, etc. are specified in Sect. 2.2.1.

When referring to beliefs or propositions (our referent (ii)), the expression "common sense" can mean: ii)a) that these beliefs or propositions are widespread within a community, society, or culture (there may or may not be epistemically good reasons for having formed or for holding these beliefs or propositions)[3]; or ii)b) that these beliefs or propositions are epistemically appropriate or sound, and easily epistemically accessible to members of the relevant epistemic community. Some common-sensical beliefs in the sense ii)b are basic causal (true) knowledge which might be pre-scientific, because there are many things that we humans can understand about the world just in virtue of materially inhabiting it.[4]

When referring to a level of judgement (our referent (iii)), the expression "common sense" can mean: iii)a) that the judgements made and the beliefs held at this level are dependent on the culture of the judge, independently of epistemically good backing, epistemic virtue or epistemic success; or iii)b) that the judgements made and the beliefs held at this level possess a certain superior degree of epistemic success.

When referring to a faculty of judgement (our referent (iv)), the expression "common sense" can mean: iv)a) that this faculty is such that leads to judgements that are dependent on the culture of the judge, independently of epistemically good backing, epistemic virtue or epistemic success; or iv)b) that this faculty is such that tends to lead to judgements with a certain superior degree of epistemic success.[5]

[3] In this sense, common sense can become loaded with scientific knowledge or with beliefs of other kinds.

[4] "Traditional" common senses can be far from the type of common sense that can be protoscientific. In modern times, our common sense has been progressively demagified and disenchanted (in the sense of Max Weber's *Entzauberung*), and loaded with modern scientific thought and knowledge. It has been largely purged of the traditional tendency to believe, and to offer, magical, mythical or otherwise non-scientific worldviews, ways to perceive the world, and explanations. Consequently, modern common sense is sociologically different from the common senses of most "traditional" societies, in that it is less inclined to appeal to magical thinking and it is generally more open to change, more open to having its beliefs falsified, is accustomed to the need for evidence, and more inclined to comply with the results of tests and to apply Ockham's razor. Epistemically, this makes modern common sense qualitatively superior to most purely "traditional" common senses, and makes it protoscientific in the sense that it is previous to science and, in conjunction with certain other provisions, gives rise to science. Some knowledge possessed by some humans in some "traditional" societies, for example related to food gathering, fishing, hunting, agriculture, or building, is highly specific and specialised, and pragmatically sophisticated and refined. This knowledge may be not widespread among most members of their respective societies, and can sometimes be the state-of-the-art knowledge in their respective societal contexts. Cases like these may be of the difficult-to-classify kinds of cases, such as the professional investigation of crimes or accidents, considered in Sect. 1.5. Senses i)b and ii)b seem to depend on the belief that neither educated guess nor science are required for the successful consideration of common-sensical matters, or that educated guess or science have already sufficiently informed common sense on the matters at hand.

[5] In humans, the common-sensical level in the sense iii)b and the common-sensical faculty in the sense iv)b are, typically, epistemically superior to wild speculation and to substandard reasoning and sensation, and inferior to educated guess and to science (cf. Bunge, 2003, p. 44). Infants and persons with mental impairments might reason beneath the level of common-sense reasoning, and

8.3 Uses of the Expression "common sense" in Philosophy

This section explores some of the most influential uses of the expression "common sense" in philosophy (those by Aristotle, Reid, and Moore).

In Aristotle's (1984, 1996 [3rd century BCE]) account of human perception and judgement, there are five special senses (smell, sight, touch, taste, and hearing). And then there is the common sense (κοινὴ αἴσθησις, *koiní aísthisis*), which is "a distinct perceptual capacity in which the five [special] senses are integrated", "a perceptual power over and above the five [special] senses which monitors their states and co-ordinates their reports" (Gregoric, 2007, pp. vii, 2). The common sense makes possible the unification of the sensations received through different senses, and allows us to perceive common sensibles (such as motion, rest, number, figure, and size (*De anima* II.5 418a17–20)) in themselves. The Aristotelian common sense is also the capacity for making various aperceptual basic judgements. These judgements allow us to recognise, and to be aware of (*De somno* 2 455a13–21; Ward Modrak, 2009, pp. 315–6), the perceptions of sensibles, and differences between different types of proper sensibles (i.e., the ones sensed via the special senses), and to "unify different types of perceptual experience" (ibid., pp. 313–8), including as manifestations of the same object (e.g., the sweetness and redness of a particular concrete bell pepper). According to Aristotle, in such cognitive grasps what exceeds the scope of the individual senses is due to the common sense (ibid.). As detailed in Sect. 8.6, what will be meant by "common sense" in this work includes, in addition to these simple judgments, judgments and beliefs that are more complex and further removed from the empirical base—including some that are products of everyday reasoning and of education—, as well as procedures, the epistemic level at which epistemically reliable common-sensical judgments occur, and the faculty that makes all those processes possible.

The Scottish school of Philosophy of Common Sense, championed among others by Reid, held that in the "perception of the average, unsophisticated man, sensations are not mere ideas or subjective impressions but carry with them the belief in corresponding qualities as belonging to external objects" (Encyclopaedia Britannica, 1998). Consistently with this, Moore, in his classic *A Defence of Common Sense* (1925), used the expression "Common Sense" (sic) to refer to the set of beliefs that we deem completely certain due to the fact that they appear to be self-evident or obvious, such as "I have hands, and they are objects that are external to my mind". He admitted that some "Common Sense beliefs" might be false, although he contends that some of them are so obvious that rejecting them would be absurd. Moore's common-sensical propositions are a part of what will be meant in this chapter by "common sense" (more specifically, they are the most cognitively accessible part of "common sense" in the sense ii)b).

---

persons with sensorial impairments might have sensorial capabilities that are inferior to the common-sensical sensorial capabilities, while educated guesses and science are done at an epistemic level that is superior to that of mere common sense.

8.4 Uses of the Expression "common sense" in Psychology

In the field of psychology, Fletcher (1984, p. 203) proposed that there are three different types of common sense, namely: "(a) […] a set of shared fundamental assumptions, (b) […] a set of maxims or shared beliefs, and (c) […] a shared way of thinking.". These, taken together, are, again, close to what will be meant in this work by "common sense", as they involve both the propositional (Fletcher's a and b, my ii) and the procedural (Fletcher's c, my i) aspects of common sense (ibid., p. 204).[6] But the notion of "common sense" is not well defined in psychology, it "is barely mentioned in psychological handbooks", and "[i]t does not seem to play much of a role in contemporary psychology" (Blankenburg, 2002, pp. 304–5).

Regarding this last point, common-sense (or folk) psychology should not be taken to necessarily include a definition of "common sense". In the expression "common-sense psychology", "common sense" does not necessarily refer to a thing defined or to be defined; rather, it refers to an unsophisticated and spontaneous level of description or explanation, that of the descriptions or explanations that a subject of study would consciously provide if asked. Common-sense psychology is the kind of psychology that describes, explains, and predicts "human thought and action in terms of beliefs, desires, hopes, fears, expectations, intentions and other so-called propositional attitudes", "belief-desire reasoning", and "emotions, traits and abilities" (Baker, 1999, p. 1).

> Commonsense psychology is what we use to explain intentional action as ordinarily described- e.g., Jack went to the store because he wanted some ice cream. Commonsense psychology also is used to explain mental states - e.g., Jill feared that she would be late because she thought that the meeting began at 4:00. Commonsense psychology is the province of everyone; we all use it all the time. (Baker, 1999, p. 1)

8.5 Uses of the Expression "common sense" in Artificial Intelligence

The expression "common sense" is also used in the field of artificial intelligence. McCarthy (1959) defined, for the case of programmes, that an *x* has "common sense" if *x* "automatically deduces for itself a sufficiently wide class of immediate consequences of anything it is told and what it already knows.". This differs from what will be meant by "common sense" in this chapter in several respects (e.g.,

[6] "This distinction between content and process is actually present in everyday usage of the term. People often say that certain beliefs are true because they are "just common sense"; that is, they are part of our common-sense store of knowledge. Conversely, the term is sometimes used to refer to the shared canons and rules that define the reasonable or correct way of thinking; for example, we occasionally urge people to use their common sense (think properly) in making decisions or judgments." (Fletcher, 1984, p. 204).

biological beings, such as humans, need not be automatic in their common-sensical inferences, and these inferences need not be deductions).

More recently, “common-sense” has been more widely defined within the field of artificial intelligence as a set of commonly-held and epistemically appropriate heuristics and modes of reasoning, and commonly-held correct knowledge. This involves both the procedural (my sense i) and the propositional (my sense ii) aspects of common sense, specifically in its senses i)b and ii)b. McCarthy himself has more recently gone down this route. In 2002 (p. 1) he wrote:

> There is common sense knowledge and common sense ability. Common sense knowledge includes facts about events occurring in time, about the effects of actions by the knower and others, about physical objects and how they are perceived, and about their properties and their relations to one another. An example is the fact that eggs contain a yolk and a white and a shell, how to recognize an egg, the effects of hard boiling them and the effects of dropping them. Common sense ability involves the use of common sense knowledge and the observation of the world to decide what to do to achieve one’s goals. [...] [W]e take common sense to include certain knowledge about the world and a certain ability to use it to infer add[i]tional facts from observations and general knowledge [...]

Another example of this is Davis (2014, p. 1), who defines common sense as “[t]he common knowledge about the world that is possessed by every schoolchild and the methods for making obvious inferences from this knowledge”.

From this use of the expression “common sense”, it follows that a system would be deemed artificially (or synthetically) intelligent if and only if it were endowed with the relevant common-sensical *faculties* (in my sense iv)b). And this degree of intelligence would be deemed a (common-sensical) *level* of “judgement” (in my sense iii)b) superior to that of non-intelligent systems or machines. So, in this modern use of “common sense” in the field of artificial intelligence, the expression is used or intended in all the above-mentioned *b* senses (i)b, ii)b, iii)b, and iv)b), which is adopted here (see Sect. 8.6) to evaluate common sense against the properties of scientificity.

8.6 The Use of the Expression “common sense” in This Work

As indicated in Sect. 8.2, common sense in the sense ii)b includes legitimate knowledge (not just mere belief), and this is the case on any account of what knowledge is in philosophy. So it is not the case that the difference between common sense and science is that science involves or leads to legitimate, reliable knowledge whilst common sense does not.

Chapter 1 indicated that common sense can be protoscientific in the sense that it is previous to science and, in conjunction with certain other provisions, gives rise to science. Clearly, “common sense” in the senses i)b, ii)b, iii)b, and iv)b refers *de dicto* to something that can be more proto-scientific than the social realities referred to when the senses i)a, ii)a, iii)a, or iv)a are used. This is because common-sensical methods and beliefs in the senses i)b and ii)b are, by this work’s definitions, objectively epistemically advisable, as opposed to common-sensical methods and beliefs in the senses i)a and ii)a, which are socially perceived as wisdom, but may lack epistemically good backing, epistemic virtue, or epistemic success, and therefore may bear very little or no relation or resemblance to scientific methods or beliefs (or even to common-sensical methods or beliefs in the senses i)b and ii)b).[7]

Methods that are socially widely-applied and beliefs that are socially widely-held either have epistemically good backings, epistemic virtue, or epistemic success, or they do not. If they do not, then they are by definition far from being scientific. If they do have epistemically good backings, epistemic virtue, or epistemic success (making them referents of the "virtuous" senses i)b and ii)b, respectively), then these methods and beliefs are either of scientific origin (and incorporated into common sense at some point in time), or originally common-sensical.[8] This later kind of methods and beliefs, i.e., those that are both originally common-sensical and epistemically well-backed, epistemically virtuous, or epistemically successful, are the most philosophically interesting contrast class for science, because they reveal how far common sense can go, epistemically-wise, by itself. Consequently, from now on we will not deal with methods and beliefs that are widespread in societies for reasons that are not sufficiently epistemically good, but we can rest assured that, by definition, they are even less scientific than epistemically good common-sensical methods and beliefs. So, from now on this chapter uses the expression "common sense" exclusively in the senses i)b, ii)b, iii)b, and iv)b. As we have seen, common sense in these senses (which are the epistemically virtuous senses) tends to possess many of the solely negatively demarcatory properties listed in Sect. 2.2.1. It also tends to lack the properties that are both negatively and positively demarcatory of science (i.e., the properties that are specific to science), as we will see in Sect. 8.8. Because the different referents of these senses (i.e., methods, beliefs resulting from the application of those methods, the epistemic level at which those procedures occur, and the epistemic faculty that makes those processes possible) can be identified by context, this work will not clarify at every step which sense is meant.

The fact that common sense includes legitimate, reliable knowledge is not sufficient for it to be science. Common sense is by definition neither science nor scientific, and virtually every philosopher has taken this position (e.g., Newton-Smith, 1981, pp. 210–212; Cromer, 1993; Wolpert, 1993; McCauley, 2000). Common sense is less epistemically potent than science (Haack, 2007, pp. 24–25, 94–109). "[T]he scientist is more careful" and "systematic" than "the common [hu]man"

---

[7] Common-sensical knowledge, including common-sensical protoscientific knowledge, includes tacit knowledge and knowledge-how, which does not even have to be propositional. Even in the case of non-propositional knowledge-how, the field could be productive of propositional knowledge, as in e.g. the so-called scientific management of workflows, also known as Taylorism.

[8] In the case of originally common-sensical beliefs that have epistemically good backing, this backing is that the beliefs in question were developed by (sufficiently) implementing common-sensical, epistemically good general practices, methods, etc. (i.e., "common sense" in our sense i)b). Many of the properties possessed by many of these methods and beliefs are specified in Sect. 2.2.1.

(Quine, 1957, pp. 5–6). "[T]he scientist is more narrowly focused on issues of truth and objectivity and, in the hope of contributing to these goals, clearer" (Hylton & Kemp, 2019).[9]

SBS "start[s] where common sense proves insufficient" (Bunge, 2003, p. 44) to deal with issues about reality that are difficult for us to accurately evaluate without a level of systematicity and criticality that is beyond common sense. And, thankfully, the limits of common sense's ability to produce knowledge might be transcended by science.

Let us see how common sense stands regarding the possession of the properties of scientificity.

8.7 The Properties of Common Sense

As already mentioned, many of the properties possessed by common sense are specified in Sect. 2.2.1. We can concede that modern common sense has all of those properties, i.e., properties 1–84.

8.8 The Properties that Common Sense (Mostly) Does Not Have

What follows is descriptive, not normative, and is not about what people think is epistemically ideal. For example, saying that common sense is more prone than science to be prey to fallacies, cognitive errors, and biases does not mean that common sense *ought* to be that way, or that there is a common-sensical rule that common-sensical epistemic agents should be that way. Rather, it simply is a description of the undesired epistemic weakness of common sense as an epistemic level in comparison with the scientific epistemic level. Arguments justifying the following are not given, because it is rather uncontroversial.

Common sense is less meticulous and rigorous than science, and not usually or generally highly precise in its formulations (lack of properties 87–89). It is not "painstaking" (lack of property 94). Common sense, especially when it is part of a culturally settled set of beliefs, is not, or is less, critical, sceptical, and examinatory than science (lack of property 118). It does not strive to subject its beliefs to rigorous testing (lack of property 122). It gives almost all weight to confirmation over falsifiability (lack of property 124). Common sense does not usually check before arguing, and often argues from unverified claims (lack of property 130). It relies on

[9] Because of the polysemy of the English expression "common sense", some of these authors' definitions of "common sense" are not exactly equivalent to mine, but are close enough to be taken as being approximately about the same referent.

anecdotal or uncritical experiential evidence, and only slowly (and sometimes with difficulty and even resistance) incorporates other kinds of evidence (lack of property 132). It relies on quantity of evidence and mostly does not consider the possibility of deficiencies in the quality of individual pieces of evidence, unless those deficiencies are apparent enough to a common-sensical thought process (i.e., when a common-sensical degree of screening is being used) (lack of property 134). Common sense usually argues in manners similar to those of science, but from a false sense of certainty, which is not even achieved by using science's safeguards (lack of property 136). Common sense is generally more prone than science to be prey to known fallacies, and cognitive errors and biases (cf., e.g., the literature on confirmation bias) (lack of properties 129 and 147). It does not use finely honed and systematic safeguards to compensate for them (lack of property 148). Common sense relies on the emotional persuasion of immediate experiences and unexamined beliefs, many of which receive additional reinforcement from being shared within an epistemic common-sensical community (lack of property 155). Common sense usually involves anachronistic thinking (i.e., thinking going back to old, refuted theories and assumptions), and it takes up falsified and rejected theories again. For example, "Aristotelian misconceptions about free fall are common among students and difficult to rectify" (Halloun & Hestenes, 1985, p. 1056) (lack of property 172).

Common sense has not adopted the general scientific method as its main strategy for dealing with its cognitive problematics (lack of property 104), and does not necessarily contain an adequate view of the general scientific method (lack of property 175). The latter is obvious from the fact that common sense is historically pre-scientific. That also implies that, historically, common sense has not always changed as a result of scientific inquiry (lack of property 176). Per se, common sense alone does not feature a special degree of disposition or concern towards properly testing theories (lack of property 206). It neither performs nor aims to perform systematic research or gathering of information for generalist purposes (lack of property 95). It does not devise rigorous tests for its propositions (lack of property 96). It does not perform planned research (lack of property 111). It does not perform procedurally valid studies (lack of property 211). Alone, it neither has enough conscious historical depth nor enough knowledge of context or of science to evaluate its ideas historically and contextually, in relation to available alternative theories (lack of property 212). It does not feature highly-specific theoretical descriptions of causal mechanisms (lack of property 179). Common-sensical theories are not usually confirmed to a degree as high as that of scientific theories (lack of property 184). This is also the case with common-sensical auxiliary hypotheses (lack of property 187). Compared to science, common sense is less bulk externally consistent with all known facts and accepted theories, especially facts discovered by science and scientific theories (lack of property 190). The knowledge that common sense provides is less systematic than scientific knowledge (lack of property 193).

Common sense neither studies the scientific domain nor has the aim of systematising general hypotheses (lack of property 195). It is not explicitly and formally peer-reviewed in written publications. Common-sensical epistemic subjects do perform peer-review of their epistemic peers' common-sensical propositions. And common sense propositions can be, and are, examined by science. But when common sense is examined in scientific publications, the examinations themselves should be scientific, not common-sensical. And common-sensical epistemic subjects, as such, are not epistemic peers of scientists qua scientists. Therefore, common sense is not subject to scientific peer review (lack of property 128). Common-sensical concern for perceived anomalies is individual or limited to individuals' immediate socialisation groups (one might ask their friends, for example, how come platypuses lay eggs if they are mammals). Concerns that go further than that would be settled by referring to science (one might consult an encyclopaedia to find out about platypuses), unless they devolve into pseudoscience or paranormal beliefs (lack of property 210).

Science is able to help common sense perform its functions (for example, keeping us alive) better than common sense alone could (lack of property 213). The cognitive and institutional mechanisms for the elimination of errors that common sense features are of a relatively epistemically weak nature (and subjected to cognitive biases, errors, and fallacies) (lack of property 214). Common sense does not feature theoretical (or programmatic) mechanisms for the elimination of anomalies. Anomalies that are perceived in common sense are better solved by science than by common sense itself. Science can accomplish this in various manners, e.g., via controlling for biases, errors, and fallacies, or by correcting conceptual frameworks (lack of property 215). Common sense has greater epistemic limitations than science. Science is required for the overcoming of those that humans can overcome. An example of these last remarks are the numerous "natural" pre-Copernican misconceptions about the Solar System detailed in the *International Handbook of Research on Conceptual Change* (2008, see especially the works by Keil and Newman, and by Chi).

Regarding all the properties in Sect. 2.2 not mentioned in this chapter, it is unclear whether common sense mostly has them.

8.9 Chapter Conclusions

By showing what kinds of properties the CF of common sense possesses, this chapter provided evidence for the appropriateness of this work's distinction between the properties that are solely negatively demarcatory of science, on the one hand, and the properties that are both negatively and positively demarcatory of science, on the other.

Common sense unsurprisingly possesses the unspecific properties of science listed in Sect. 2.2.1, but does not clearly have any, and clearly lacks at least 36, of the specific properties of scientificity featured in the demarcatory list. This yields a score of −36 for common sense. This (possibly provisional) tally is strongly against the scientificity of common sense, and is the best explanation for the fact that common sense is commonly deemed non-scientific. This is yet one more case in which my demarcatory procedure demarcates in a manner that is equivalent to that of the current international philosophical and scientific consensus regarding the extension of the term "science".

References

Aristotle (1984 [3rd century BCE]). De somno. In Barnes, J. (Ed.), *The complete works of Aristotle: The revised Oxford translation*. Princeton University Press.
Aristotle (1996 [3rd century BCE]). *De anima*, Shields, C. J. (trans.). Oxford: Oxford University Press.
Baker, L. R. (1999). What is this thing called 'Commonsense Psychology'? *Philosophical Explorations, 2*(1), 3–19.
Blankenburg, W. (2002). First steps toward a psychopathology of "Common Sense". *Philosophy, Psychiatry, & Psychology, 8*(4), 303–315.
Bunge, M. A. (2003). *Philosophical dictionary* (Enlarged ed.). Prometheus Books.
Chi, M. (2008). Three types of conceptual change: Belief revision, mental model transformation, and categorical shift. In S. Vosniadou (Ed.), *International handbook of research on conceptual change* (pp. 61–82). New York and Abingdon-on-Thames: Routledge.
Cromer, A. (1993). Uncommon sense: The heretical nature of science. *Science, 265*, 688.
Davis, E. (2014). *Representations of commonsense knowledge*. Elsevier Science.
Encyclopaedia Britannica. (1998). Philosophy of common sense. https://www.britannica.com/topic/philosophy-of-common-sense
Fletcher, G. J. O. (1984). Psychology and common sense. *American Psychologist, 39*(3), 203–213.
Gregoric, P. (2007). *Aristotle on common sense*. Oxford University Press.
Haack, S. (2007). *Defending science—within reason: Between scientism and cynicism*. Prometheus Books.
Halloun, I. B., & Hestenes, D. (1985). Common sense concepts about motion. *American Journal of Physics, 53*(11), 1056–1065.
Hylton, P., & Kemp, G. (2019). Willard van Orman Quine. In E. N. Zalta (Ed.), *The Stanford Encyclopedia of Philosophy* (Spring 2019 ed.). Stanford University. https://plato.stanford.edu/archives/spr2019/entries/quine/
Keil, F. C., & Newman, G. E. (2008). Two Tales of conceptual change: What changes and what remains the same. In S. Vosniadou (Ed.), *International handbook of research on conceptual change* (pp. 83–101). New York and Abingdon-on-Thames: Routledge.
McCarthy, J. (1959). *Programs with common sense*. http://www-formal.stanford.edu/jmc/mcc59.pdf
McCarthy, J. (2002). *What is common sense*. http://jmc.stanford.edu/articles/commonsense/commonsense.pdf
McCauley, R. N. (2000). The naturalness of religion and the unnaturalness of science. In F. C. Keil & R. A. Wilson (Eds.), *Explanation and cognition* (pp. 61–85). The MIT Press.
Moore, G. E. (1925). A Defence of common sense. In J. H. Muirhead (Ed.), *Contemporary British philosophy (2nd series)* (pp. 32–59).
Newton-Smith, W. (1981). *The rationality of science*. Routledge.
Quine, W. V. O. (1957). The scope and language of science. *British Journal for the Philosophy of Science, 8*, 1–17.
Ward Modrak, D. K. (2009). Sensation and desire. In G. Anagnostopoulos (Ed.), *A companion to Aristotle*. Wiley-Blackwell.
Wolpert, L. (1993). *The unnatural nature of science: Why science does not make (common) sense*. Harvard University Press.

Chapter 9
The Properties of Art

Abstract This chapter provides a brief definition and a description of/introduction to art. It then meta-analytically shows that the demarcation criteria and the scoring method proposed in Chap. 2 demarcate art as non-scientific, in agreement with the current philosophical and scientific consensus.

Keywords Art · Demarcation · Science · Non-science · Scientificity

9.1 Introduction

Virtually everyone believes that art is non-scientific. This section gives a working definition of art. Section 9.2 lists properties of science that art has. Section 9.3 lists properties of science that art lacks. As the conclusion of this chapter, Sect. 9.4 argues that the demarcation criteria listed in Sect. 2.2 demarcate art as non-scientific, and that the fact that virtually everyone believes that art is non-scientific is explained by the fact that it does not have enough of the properties of science listed in Chap. 2.

The term "art" is difficult to define. Conventionalist, procedural, institutional, historical, functional, and hybrid definitions have been given in a massive amount of literature (see Davies, 1991; Abell, 2012; Adajian, 2018). Because I cannot define art in passing, the best course of action here is to adopt the following very general and loose, working definition of art, which, I submit, is sufficiently broad that it accommodates all clear cases of art: art is the activity of consciously producing works, performances, features, or situations that are to be appreciated primarily for (or via) their aesthetic properties (e.g., beauty), emotional, or experiential power, or that are intended to convey meaning or a message or suscitate non-purely intellectual appreciation, reflection, or insight by a collection of means most often chiefly including non-literal means. Those works, performances, or situations are also called art. (Authors disagreeing with this definition of art would have to modify my analysis accordingly. But in any case, my determination on whether or not art possesses a particular property does not depend on this definition of art, and none of my adjudications of properties is at issue among the authors who argue about what art

D. Fernandez-Beanato, *A Working Scientific Demarcation*, Synthese Library 519, https://doi.org/10.1007/978-3-031-99005-2_9

is.) In this way, art covers an entire range from artefacts that are merely pretty to things that make people feel and reflect. "[A]rtworks can have considerable moral and political as well as aesthetic power" (Adajian, 2018).

As shown below, it is a truism that art is non-scientific.[1] The purpose of this discussion is not to contribute anything new to the question as to whether art is scientific. The concern of this chapter is what properties of science art has and what properties of science it lacks. Art is considered to help make the case that the SD proposed here is not only successful in demarcating science from pseudoscience, but also in demarcating science from other non-scientific fields of activity, such as art and religion (which is covered in Chap. 10). I take the properties of science listed in Sect. 2.2, and show that art possesses many of them. This provides: a) a rational reconstruction (and therefore an explanation) of *why* art is non-scientific, b) further evidence of the extensional adequacy of this work, c) further evidence of this work's explanatory power, and d) further evidence of the fact that the proposed set of criteria demarcates science appropriately.

The differences between art and science are deep. Art, as a field of activity, is not, even purportedly, primarily aimed at gaining knowledge, so it is not even a CF according to our definition of "cognitive field" (cf. Note 11 of Chap. 1). This fact alone is sufficient to put art outside of the superdomain of science. Therefore, if our main hypothesis is true, then the demarcatory list should clearly demarcate art as non-scientific. This chapter's analysis should also help us get an enhanced understanding of what art is, how it differs from science, and the reasons why it is not scientific.

The method adopted in this chapter is to draw from works on what art is (Davies, 1991; Abell, 2012; Adajian, 2018), and then to assess which of the properties of science art possesses and which it lacks, in the process using works of some of the authors who have found the most commonalities between art and science (Goodman, 1976; Elgin, 1991, 2010) to enrich the analysis and ascertain what uncontested differences remain between art and science.

Let us see how art stands regarding the possession of our scientificity properties.

9.2 The Properties of Art

Arguments justifying most of the following claims regarding the possession or lack of properties by art are not given, because these judgements are fairly clear and non-controversial.

Art has the following properties: it does not propose or use ad hoc definitions or hypotheses to evade theoretical falsification (properties 9–10), it does not use unfounded immunisations (property 12), it does not typically refer to alleged

[1] Some artistic projects are tied to goals other than purely artistic ones. Nothing in this chapter rules out that one and the same thing (e.g., a drawing of an animal) can be, or be legitimately deemed, both scientific and artistic at the same time, but always in different manners, which are respectively characterised by the differences between science and art that are outlined in this chapter.

phenomena at the very limit of detection (property 61), it does not usually attack logic (property 66), it is obviously a creative activity (property 68), it can be meticulous and painstaking (properties 85 and 94), it does not typically suppress data (property 98), it usually avoids both a belief in personal authority and dogmatism (properties 108 and 117), it can incorporate a critical, sceptical and examinatory attitude (property 118), it does not argue from spurious similarity to science (which is why it is not a pseudoscience) (property 136), it does not usually deem tradition as deserving to be worshipped (property 160), it does not feature dubious forms of science-sounding mechanics (property 180), it does not typically have or use built-in subterfuge (property 185), its products do not feature unsolved cognitive problems or anomalies (property 191), and it can be very open-minded (property 199).

9.3 The Properties that Art (Mostly) Does Not Have

It is fairly clear that art itself does not produce theoretical findings (lack of property 1). Even when it has a message, art is not essentially objective (lack of property 2), because it depends on the whims of artists. Artistic freedom is bigger and wider than epistemic freedom: whilst, on the one hand, artists can (within certain constraints) do and say whatever they like or want, on the other hand people intending to believe only what is most probably true, or truthlike, or reliable, or warranted, or even useful cannot just believe anything they would like to believe.

Art is not, essentially, propositionally factive (lack of property 4). It does not feature confirmations and disconfirmations (lack of property 7). It is not essential to art to look for correlations (lack of property 23). Usually, it does not explicitly feature cognitive problems (lack of properties 40 and 81, the latter because it presupposes the existence of cognitive problems). Although, as indicated in Sect. 9.1, there are many accounts and theories on art, these theories themselves are not art; art itself offers no explicit, detailed, and unambiguous theories or other types of accounts, and has very low explanatory power regarding observable phenomena (lack of properties 42, 169, and 212, the latter two because they require the existence of theories). It does not postdict (lack of property 44). It does not usually propose causal mechanisms (lack of properties 48 and 179). It is not eminently useful for forming true, truthlike, or reliable beliefs (lack of property 50). Because it is mostly non-propositional, it is mostly non-testable (lack of property 51), and mostly non-falsifiable (lack of property 54). Because art offers no theories, it lacks connected doctrines (lack of property 57). It does not usually set phenomena within a unified framework (lack of property 60). Because it is mostly non-propositional, it is mostly not empirically fallible/provisional/tentative (lack of property 63). Art is not essentially rational; it does not follow logical rules, or concepts, judgements, and arguments organised in systems (lack of property 65). It is more chaotic than that. Epistemically, it is not very fruitful (lack of property 82), shows no substantial promise of fruitful development (lack of property 83), and it is not very progressive (lack of property 84). It is not methodically rigorous (lack of property 88).

Art is not propositionally precise (lack of property 89), or propositionally clear (lack of property 92). It is not propositionally clear in part because

> [a]esthetic interpretations, unlike scientific ones, are endlessly contestable [... (,)] there is no consensus about how to interpret works of art [... (and)] the reasons supporting an interpretation of a work of art are apt to be inconclusive. [...] There is far less agreement [than in science] about the adequacy of [...] interpretations [...] (Elgin, 2010, pp. 15–6)

Between science and art there is

> a divergence in [...] ends. A work of art always remains open for interpretation, drawing the spectator into the shape of the artist's visualization, but without being able to exert fixed control over the feelings it induces. There is always room for the beholder's share. Scientists may wish to engage the reader or spectator in a wonderful journey of imaginative visualization, but in the final analysis they wish to communicate an interpretation that embodies [...] content in an unambiguous way. (Kemp, 2005, p. 309)

None of this should be taken to mean that art cannot promote understandings, impart information, or convey messages, even profound or important ones. Unlike other fields of conscious and purposeful activity (e.g., cooking for oneself), art is regularly used to convey cognitive meanings or to transmit messages. The transmission or communication of cognitive meanings or messages can even be the primary purpose or the biggest, sometimes unintended, impact of a particular work of art or artistic situation or performance. "[T]he issue is not whether, but *how* the arts function cognitively" (Elgin, 1991, p. 196, emphasis added). When art conveys cognitive meanings or messages, these are not as explicit, literal, or precise as messages communicated in scientific publications or even in everyday speech. In this way, "the meanings [... (that art works, performances, or situations)] have differ [...] from the meanings of newspaper reports, conversations, instruction manuals and other things that are not artworks" (Abell, 2012, p. 673). Picasso's *Guernica* can give an observer a (more or less) clear idea of chaos, disarray, and suffering, but a single page of a good history book could be much more (epistemically) effective in precisely telling the observer much more about the 1937 bombing of the town of Guernica and its results. This pertains to cognitive understanding and not necessarily to empathetic understanding, which is more suited to be of the province of art: when art does transmit messages, those messages themselves are often enmeshed in, or intertwined with, emotional meaning. This is why artistically-delivered messages can affect people deeply, sometimes in ways that are even more effective or efficient, in some respects, than those of science. In this way, art conflates cognitive meaning with emotional meaning (e.g., in Ut's *The Terror of War*), which is unadvisable in science and not always helpful in everyday communication. But simply believing the proposition 'War is horrible' might not be as effective or efficient, in some respects, as internalising a more intense corresponding feeling.

Art does not systematically gather information (lack of property 95). It does not perform carefully controlled, demanding, and precise observation/study, does not devise rigorous tests for propositions, and does not convince by appeal to evidence, or by arguments based upon logical and/or mathematical reasoning. It does not need to abide by evidence, can be indifferent to facts without detriment, and does not feature experiments (lack of property 96). It uses handpicked ("cherry-picked") examples (lack of property 97) (e.g., that of Margarita Teresa in the first interpretation in the discussion on Velázquez's *Infanta Margarita Teresa in a Blue Dress* in Elgin, 2010, p. 15). Art does not feature screening and sorting of putative facts that is scientifically appropriate to separate pseudofacts from facts (lack of property 99). It does not present or incorporate coherent evidence/proof (lack of property 100). It does not mainly use careful scrutiny (lack of property 102). Its methodics do not mainly consist of scientific methods (lack of property 103). It has not adopted the general scientific method (lack of property 104). Its (implicit) descriptions do not typically have to be independently recognised as correct (lack of property 105). It does not feature theoretically-directed research or investigation (lack of property 111). It does not usually use control groups (lack of property 113). It does not treat reality systemically and does not look for systemicity (lack of property 116). It does not tend to revise or check in light of new evidence (lack of property 120). It does not test (lack of properties 122 and 207). Art qua art is obviously not subject to scientific peer review (lack of property 128). It uses metaphors (lack of property 139). It does not use finely honed and systematic safeguards to compensate for cognitive biases (lack of property 148). It is indifferent to criteria of valid evidence (lack of property 154). It relies on emotional impact and appeals to emotion or sentiment, and it ignores scientific literature (lack of property 155).

It is not systematic, does not primarily aim at the pursuit of knowledge, and does not provide systematic knowledge (lack of property 178). Art is not legal (lack of property 181) and does not usually propose explanations (lack of property 182). Because art offers no explicit, detailed, and unambiguous theories, it lacks theoretical confirmation (lack of property 184), and auxiliary hypotheses (lack of properties 58 and 186–187). It does not intend to search, discover, or use regularities, and it neither aims to generalise information into hypotheses nor to systematise hypotheses into theories (lack of property 195).

Art does not perform procedurally valid studies (lack of property 211). It does not feature cognitive, methodological, institutional, theoretical, or programmatic mechanisms for the elimination of cognitive or epistemic errors (lack of properties 214–215). And it is not epistemically self-corrective (lack of property 217).

Because art is not primarily aimed at the pursuit of knowledge, it is not a research field. Because of this and of the fact that it offers no explicit, detailed, and unambiguous theories, art as a field is not constituted, even partially, by a community of theorists (there are theorists of art, but their theorising itself is not art). Because it lacks a community of theorists (whose possession is presupposed in Chap. 2 as a precondition for many of our properties, which therefore art lacks) and it is not primarily aimed at the pursuit of knowledge, art is not concerned with reducing the number of unexplained phenomena (lack of properties 41, 73–76, 195, 201, 204–208, 210, and 219).

Regarding all the properties in Sect. 2.2 not mentioned in this chapter, it is unclear whether art mostly has them. The heterogeneity of art makes very difficult to adjudicate the possession or lack of many of our properties to it. Take for example our property 3: is art concerned with truth and falsity? Artworks intended to send propositional messages appear, at least at some level, to be so, but other types of art, such as abstract art, are clearly not necessarily concerned with truth and falsity. Consider also property 109: does art recognise the epistemic authority of science or of the scientific community? Some art might, but it will depend on the artist or on the piece. What about our property 183? Is art highly empirically adequate? Some art is, in some respects, but obviously not all of it, not even all representational art, is. Matisse's semi-abstract *The Yellow Curtain* is based on an experience of a view from a window at the artist's home in Issy-les-Moulineaux, but there is no clear commitment to a correspondence to the real view that originally inspired the artist. Not even highly realistic landscape representations, such as Vermeer's *View of Delft*, provide any guarantee or even an implicit assertion that they are faithful representations of a real landscape. Such is the multifacetedness of art. Consequently, I do not adjudicate (i.e., I remain neutral) on art's possession or lack of these properties. In this way, my analysis uses only a minimal set of assumptions in this regard, and is more charitable towards the degree of scientificity of art in that I do not deny that art has any of these properties.

9.4 Chapter Conclusions

Art lacks at least 72 of the properties of scientificity featured in the demarcatory list, while having at least 12 of the specific ones. This yields a score of −60 for art. This (possibly provisional) tally is strongly against the scientificity of art, and is the best explanation for the fact that art is commonly deemed non-scientific. This, once again, demarcates in a manner that is equivalent to that of the current international philosophical and scientific consensus regarding the extension of the term "science".

References

Abell, C. (2012). Art: What it is and why it matters. *Philosophy and Phenomenological Research, 85*, 671–691.

Adajian, T. (2018). The definition of art. In E. N. Zalta (Ed.), *The Stanford Encyclopedia of Philosophy* (Fall 2018 ed.). Stanford University. https://plato.stanford.edu/archives/fall2018/entries/art-definition/

Davies, S. (1991). *Definitions of art*. Cornell University Press.

Elgin, C. Z. (1991). Understanding: Art and science. In *Midwest studies in philosophy, XVI*.

Elgin, C. Z. (2010). Telling instances. In R. Frigg & M. C. Hunter (Eds.), *Beyond mimesis and convention: Representation in art and science*. Springer.

Goodman, N. (1976). *Languages of art. An approach to a theory of symbols* (2nd ed.). Hackett.

Kemp, M. (2005). From science in art to the art of science. *Nature, 434*, 308–309.

Chapter 10
The Properties of Religion

Abstract This chapter provides a brief definition and a description of/introduction to religion. It then meta-analytically shows that the demarcation criteria and the scoring method proposed in Chap. 2 demarcate religion as non-scientific, in agreement with the current philosophical and scientific consensus.

Keywords Religion · Scientific demarcation · Non-science · Pseudoscience · Scientificity

10.1 Introduction

Virtually everyone believes that religion is non-scientific. This section provides a working definition of the term "religion". Section 10.2 lists properties of science that religion has. Section 10.3 lists properties of science that religion lacks. As the conclusion of this chapter, Sect. 10.4 argues that the demarcation criteria listed in Sect. 2.2 demarcate religion as non-scientific, and that the fact that virtually everyone believes that religion is non-scientific is explained by the fact that it does not have enough of the properties of science listed in Chap. 2.

The purpose of this chapter's discussion is not to contribute anything new to the question as to whether religion is scientific. The concern of this chapter is what properties of science religion has and what properties of science it lacks. Religion is, at least partially, purportedly aimed at gaining certain kinds of knowledge, and therefore it is a CF according to our definition of "cognitive field" (cf. Note 11 of Chap. 1). But, as we will see in this chapter, the differences between religion and science are deep. Therefore, if our main hypothesis is true, then the demarcatory list should clearly demarcate religion as non-scientific. This chapter's analysis should also illuminate us by helping us to get an enhanced understanding of what religion is, how it differs from science, and the reasons why it is not scientific.

Religion is considered in this work to show that the SD proposed here is not only successful in demarcating science from pseudoscience, but also in demarcating science from non-scientific fields, such as art (which is covered in Chap. 9) and

D. Fernandez-Beanato, *A Working Scientific Demarcation*, Synthese Library 519, https://doi.org/10.1007/978-3-031-99005-2_10

religion. Another motivation for the consideration of religion here is that, according to the Pew Research Center (2012), approximately 84% of all humans declare themselves to be religious believers, or at least to be affiliated with a religion. Thus, not only for epistemological but also for social reasons, an SD methodology and its resulting taxonomy would be more complete and would be more fit for purpose if they dealt with the question of what is the appropriate categorial placement of supernatural religion (and of theology) within said taxonomy. Therefore, considerations about supernatural religion and theology have a place in my analysis on SD.[1]

This chapter takes the properties of science listed in Chap. 2, and shows that religion possesses some, and lacks many of them. This provides: a) a rational reconstruction (and therefore an explanation) of *why* (everyone believes that) religion is non-scientific, b) further evidence of the extensional adequacy of this work, c) further evidence of this work's explanatory power, and d) further evidence of the fact that the proposed set of criteria demarcates science appropriately.

The academic literature on science and religion is extensive (see e.g. Ferngren, 2000; van Huyssteen, 2003; Clayton & Simpson, 2006; Dixon, 2008; Southgate et al., 2011; De Cruz, 2022).

Let us start by defining the term "religion". There is no generally accepted definition of the term (Pollack, 1995, p. 163). So-called philosophical, substantive, hermeneutical, and explanatory definitions have been proposed (Pollack, 1995).

Note that

> [t]he concept of religion [... (but not religion itself)] is the product of, among other influences, the Age of Exploration and Empire (fifteenth and sixteenth centuries), the Protestant Reformation (sixteenth century), and the Augustan Age (eighteenth century). [...] [T]he premodern etymological antecedents of the modern word *religion* (e.g., Latin *religio*) generally mean something like "the pious [...] rites of worship," not what the modern word means. The practice of translating the premodern terms as *religion*, therefore, often misleads. Non-Western cultures, furthermore, did not have terms with anything like the same

[1] Except for this note, this chapter analyses religion and not theology (which has some more of the properties of science). But let us review here some of the main problems with classifying any theology—including a non-revealed one, such as a natural theology—as broadly scientific. An important objection is that of i) the low degree of specificity of the claims connecting theological propositions to natural-world propositions. Science, confronted with questions, comes up with specific answers (properties 46 and 179). Also, ii) because of i), theology lacks the ability of making novel testable/contrastable predictions (a trait that is present in science and that is related to fruitfulness (property 82)), and (iii) theological statements are underdetermined by the available evidence (or facts, or data) in a higher degree than the statements of SBS are. Therefore, the argument goes, (iv) theology is less, or not at all, falsifiable (property 54), while broadly scientific theories, hypotheses, and particular statements are. Points (i) and (iv) are the reasons, for example, behind the fact that after more than a millennium and a half, Orthodox and Roman Catholics cannot agree on whether the Holy Spirit "proceeds" only from God the Father, or from both the Father and the Son, because there is simply no way to indicate how any of those cases being true would have any effect on any agreeable-upon empirical proposition, so neither of the claims can be falsified or encounter anomalies. This takes us to what is apparently a further differentiating characteristic: (v) science progresses (property 84) in a way and to a degree that is unparalleled in any theology.

> connotations or semantic scope. [...] [R]eligion [... (is)] a general category identifying one aspect of human intellectual, emotional, and social life. (Bagger, 2003, pp. 724–5)

All clear cases of religion are accommodated if we understand a religion as a system of beliefs that aims at transcendentally relating some conscious sentient beings (in all cases that we know of, humans) to alleged supernatural (i.e., above the natural) or spiritual realities or entities (here, the term "spiritual" excludes the senses "of or relating or pertaining to the mind" and "of or relating or pertaining to emotions"[2]). Although the most populous human religions, Christianity and Islam, are strongly theistic, religions are not necessarily theistic. For example, gods do not play a religious role in some varieties of Buddhism (Glasenapp, 1966) and Jainism. Also, some religions, for example Buddhism, can be stripped of religious content and converted into non-religious philosophies of life. In this way, the spectrum of these movements falls on both sides of the line demarcating religion from non-religion, so that there are, for example, non-religious Buddhists.

Authors disagreeing with this definition of religion would have to modify my analysis accordingly. But in any case, my determination on whether or not religion possesses a particular property does not depend on this definition of religion, and none of my adjudications of properties is at issue among the authors who argue about what religion is.

Let us see how religion stands regarding the possession of our scientificity properties.

10.2 The Properties of Religion

Arguments justifying most of the following claims regarding the possession or lack of properties by religion are not given here, because these judgements are rather clear and non-controversial.

We concede that religion has the following properties: it looks for correlations (property 23), it has a particular domain or universe of discourse, in the best cases its discourse and the terms it uses in it have meaning, and those terms have precise and unambiguous definitions (properties 34–36); it presupposes philosophical background assumptions, at least implicitly, for example, its general outlook or philosophical background includes the epistemological principle that the world can be known objectively, at least partially and gradually (properties 37–38). In the best cases, religion can view science as organised activity which aims at least in part at factually truthlike descriptions (property 39). Religions offer accounts. If we use the term "explanation" in a non-necessarily factive sense, religion has the capability to be used to explain; it has the power to explain observable phenomena, especially

[2] This clarification is necessary because in some European languages a single word—e.g., the English "spirit", the Spanish "espíritu", the German "Geist"—can be used to (try to) refer both to embodied minds and to alleged disembodied spirits.

intended data, and it can sometimes explain all the important data. Consequences of religious theories are themselves explanatory. And religion has explained new classes of facts (property 42). Religion features proper, comprehensive (consilient) but yet sufficiently specific theories or at least "models". Religious theories have broad scopes (property 46). Many religious claims are general. Because of that, religion seems to have the capability to bring "order to phenomena that [...] would [otherwise] be individually isolated and, as a set, confused" (property 47). Religion can explain using mechanisms (property 48). Religion is eminently communicable and expressible, sometimes with the exception of certain points that are said to be ineffable or mysterious (e.g., the Trinity in Christianity). Religious language can communicate information (property 49). Religion features theoretical boundary conditions (property 53) and its beliefs are transparent/public and open for all to scrutinize (except for mystery religions) (property 55). Developed religions are "not a gerrymandered patchwork but a coherent whole [...] setting the phenomena within a unified framework" (quote: Kitcher, 1982, pp. 30–54) (property 60), and, because they aim to explain the observable world in which we live, most of the phenomena that they cover are not at the very limit of detection (property 61). Religion is not omniscient, nor its theoreticians or practitioners claim that it is so (property 62). It is parsimonious/unificatory in its problem-solving strategies, consisting "of just one problem-solving strategy, or a small family of problem-solving strategies, that can be applied to a wide range of problems. The theory [...] is able to encompass more and more problem areas". "Problems are solved by invoking the same pattern of reasoning, or *problem-solving strategy*, again and again". "*Unification* is the result of applying a small family of problem-solving strategies to a broad class of cases." (quote: Kitcher, 1982, pp. 30–54) (property 81).

10.3 The Properties That Religion (Mostly) Does Not Have

It is fairly clear that religion does not produce theoretical findings (lack of property 1), and that it does not feature confirmations and disconfirmations of its most basic/essential and central synthetic beliefs (lack of property 7).[3] It uses unfounded immunisations, for example in the case of official dogmas or *articuli fidei* (e.g., the Immaculate Conception and the Assumption of Mary in Roman Catholicism)[4] (lack

[3] Disconfirmations of important, even central beliefs happen sometimes, for example when religions dare to make predictions that are chronologically specific. For example, the passage of time forced Jehovah's Witnesses to modify their claim that the believers alive in 1914 would never completely die out, a claim that had been central (but not theologically basic/essential) to them for decades.

[4] Some religions are less dogmatic than others. Some of the less dogmatic religious schools adopt a more "whatever works" attitude about their main goals. As indicated above, some religions, for example Buddhism, can be stripped of (religious) dogmas and other religious contents and converted into non-religious philosophies of life.

of property 12). It does not check for results or effectiveness in order to ascertain whether believers should continue to believe (lack of properties 19–20).

Religion is not based on "the logic of inconclusive justification" (lack of property 30).

It lacks a fund of knowledge which is a collection of up-to-date, testworthy and testable (and at least in part tested and confirmed) (yet corrigible) data, hypotheses, and theories which are compatible with its general philosophical views, as well as of other information previously incorporated, and of reasonably effective research methods (lack of property 41). It features no particular predictive power (lack of properties 45, 82, and 184). Religion is not directly useful for developing technology (lack of property 50). It is not testable (lack of property 51), its propositions are not intersubjectively confirmable or disconfirmable (lack of property 52), and it is not in-principle falsifiable (lack of property 54). The experiences on which it is based are not in-principle repeatable/replicable by every intellectually-capable adult (lack of property 56). It does not present its most basic/essential and central synthetic beliefs as fallible/provisional/tentative (lack of property 63). Religious accounts are not intersubjectively valid, i.e., they lack publicly accessible supporting evidence. They do not result in publicly verifiable specifically religious knowledge (lack of property 64).

Religions are largely unconcerned with generating correct predictions (and when they have done it they have failed catastrophically, e.g., the Millerite Great Disappointment of 1844, and the eschatological predictions by Jehovah's Witnesses during the Twentieth century. When this happens, religious groups usually use spurious strategies to make disconfirmations appear as confirmations, e.g. invisible events in Heaven described by Adventists and Bible Students) (lack of property 76). Religion lacks fruitfulness of research findings of "new phenomena or previously unnoted relationships among those already known" (quote: Kuhn, 1977, p. 322) (lack of property 82). Religion does not promise further fruitful epistemic or technological development (lack of property 83). It does not progress in "learning about [...] processes under study," (quote: Coker, 2001, p. 6) (lack of property 84).

Its methodics do not consist exclusively of scrutable (checkable, analysable), criticisable, justifiable, explainable, testable and historically correctable procedures which are appropriate scientific methods and techniques (lack of property 103). Religion has not adopted the general scientific method as its main strategy for dealing with its cognitive problematics (lack of property 104). Its descriptions and explanations cannot be independently recognised as correct (lack of property 105). Religions often accept personal epistemic authority other than the one derived from knowledge, training, experience, study, and reflection. They often claim special insights for special "initiated" people other than the insight coming from knowledge, training, experience, study, and reflection. This is seen in the religious acceptance of revelations, visions, familial links, membership in dynasties, initiations, consecrations, apostolic successions, etc. as legitimate sources of, and indications of the possession of, epistemic authority or insight (lack of property 108).

Religion neither studies the scientific domain nor has the aim of systematising general hypotheses (lack of property 195). It is clear that religion does not necessarily contain an adequate view of the scientific method. This is obvious by the fact that religion is historically pre-scientific (i.e., it appeared before science) (lack of property 175). That also implies that, historically, religion has not always changed as a result of scientific inquiry. Although the extension of what religion(s) aspire to explain has been progressively reduced due to the advancement of scientific knowledge, religion(s) themselves do not typically change, nor have typically changed, as a result of research results obtained in and through scientific inquiry (formal or empirical) (lack of property 176). Per se, religion alone does not feature a high degree of disposition, willingness, and concern towards properly testing, confirming or disconfirming theories. They are not *willing to test* and do not put their claims to meaningful tests such as careful, methodical experiments (lack of property 206). Therefore, religions cannot feature high degrees of disposition of their communities to comply with the results of such tests (lack of property 207). Religion does not perform planned research, neither performs nor aims to perform systematic research or gathering of information for generalist purposes, it mostly does not perform research at all, and does not feature theoretically-directed investigation (lack of properties 93 and 111). It does not devise rigorous tests for its propositions (lack of property 96). Religion does not perform procedurally valid studies (lack of property 211). It does not investigate or test alternative possibilities. It does not have enough knowledge of context or of science to evaluate its ideas contextually, in relation to available alternative theories (although theology might) (lack of properties 152 and 212). Religions have not been more epistemically progressive than alternative theories over a long period of time (lack of property 213). Anomalies that are perceived in religion are better and more simply solved or disolved by science than by religion or theology (e.g. of dissolution: there is no Problem of Evil if there is no God) (lack of property 216). Religion has greater epistemic limitations than science. Religions are often based on some alleged paranormal epistemic super-abilities of an elite (patriarchs, messiahs, prophets, gurus, "enlightened" or "spiritually anointed" people, channelers, contactees, etc.) (lack of property 220).

Religion does not feature highly-specific theoretical descriptions of causal mechanisms (lack of property 179). Religious theories are not usually confirmed to a degree as high as that of scientific theories (lack of property 184). Religion does not check for variation of variables (lack of property 112). It neither uses control groups nor blind (masked) testing (lack of properties 113–114). It does not get independent confirmation of religious alleged facts (lack of property 115). Religious auxiliary hypotheses are not usually confirmed to a degree as high as that of scientific auxiliary hypotheses (lack of property 187). Religion is not contiguous with any scientific research field, in that it does not share items in its general outlooks, formal backgrounds, specific backgrounds, specific funds of knowledge, aims, and methodics with any science; and in that its domain is not included in the domain of any science and it is not the case that each member in the domain of religion or of any science is a component of a system in the domain of the other (lack of property 189).

Religion does not avoid dogmatism or dogmatic assertions (lack of property 117). Religion is not critical and sceptical, or is less critical and sceptical than science, which can be especially seen in the existence of religious dogmas (lack of property 118). Religions have *built-in subterfuge* in that they are not open to falsifications or disconfirmations of their fundamental tenets. Ecumenical and inter-religious dialogue and efforts almost always implicitly presuppose that converting from one religion to another, or changing the basic tenets of the participant religions, is not an acceptable and open possibility for each religion's adherents (lack of property 185). Religion does not tend to revise or check its proposals in light of new evidence or criticism (and per se, religion alone does not tend to revise or check its proposals in order to strengthen their scientific content). Some religions claim to be open to scientific falsification (e.g., the Gelug or "Yellow Hat" (also known as "New Kadam") school of Tibetan Buddhism, at least according to its foremost spiritual leader, Tenzin Gyatso (the 14th Dalai Lama)). But these claims are substantive (in the best cases) only about non-essential matters; they are vacuous on essential matters because religions' distinctive basic tenets are not usually directly scientifically falsifiable, with the consequence that religions will not consider them to have been falsified, and, consequently, will not revise them. Reincarnation is a good example of this: science does not accept reincarnation, although Gelug Buddhists do; and it is not clear what the Dalai Lama would accept as a falsification of reincarnation (lack of property 120). Because religions are not open to falsifications or disconfirmations of their fundamental tenets, their general outlook does not include the ethos of the free search for truth (lack of property 194). Not being open to falsifications or disconfirmations of their fundamental tenets implies a lack of epistemic humility (the psychological or at least sociological awareness or at least recognition of the fact that one might be incorrect or mistaken) and of open-mindedness (lack of properties 198–199). As a result, religions also lack research communities featuring sufficiently deep and free traditions of inquiry (lack of property 201). Religions lack organised scepticism allowing detached scrutiny of beliefs that are dearly held. This sometimes brings science into conflicts with religions and other ideologies (lack of property 205).

Religious propositions can be, and are, examined by science. But when religion is examined in scientific publications (e.g., sociological publications), the examinations themselves should be scientific, not religious. And religiously epistemic subjects, as such, are not epistemic peers of scientists qua scientists. Therefore, religion is not formally subject to scientific peer review in scientific publications (lack of property 128). Religion usually argues from a false sense of certainty, whereas certainty is not even achieved by using science's safeguards (lack of property 136). Religion does not avoid emotional attachment to its ideas (lack of property 121). Religion often relies on emotional impact and appeals to emotion or sentiment (lack of property 155). It relies heavily on subjective validation (lack of property 156). Religion is not "continually striving to find ways of subjecting" its "most cherished beliefs to rigorous testing" (quote: Lilienfeld et al., 2012, p. 12) (lack of property 122). Religion gives excessive weight, in some cases almost all weight, to confirmation over falsifiability or over falsification (lack of property 124). Religion does "research by exegesis" (lack of property 137), uses metaphors (lack of property 139), and improperly restricts acceptance of data interpretations to certain particular interpretations (lack of property 142). Religion does not use finely honed and systematic safeguards to compensate for known cognitive errors and biases (lack of property 148). Religion often involves anachronistic thinking (i.e., thinking going back to old, refuted theories and assumptions), and it takes up falsified and rejected theories again (lack of property 172).

Regarding all the properties in Sect. 2.2 not mentioned in this chapter, it is unclear whether religion mostly has them. Take for example our property 25: does religion appeal to unfounded non-mysteries? When religion starts, it provides answers to what are, contextually and at the time, legitimate mysteries. The origin of the Universe continues to be mysterious to some degree even today. At the same time, the more we scientifically learn, the less mysterious these matters seem to be. So it is not clear that religion has completely ceased to appeal to, and try to provide answers about, legitimate mysteries. Consider also our property 157: does religion depend on arbitrary conventions of human culture? Some of the questions that religions try to answer seem to be quite universal to humans, whilst other, more specific aspects of religion (e.g., the properties of Vishṇu's (or Krishṇa's) Vishvarūpa (universal form)) are more conventional. What about our property 183? Is religion highly empirically adequate? Religion has at times been adequate, and when it has ceased to be so (due to new discoveries), it has at times resisted change and at other times strived to accommodate itself. Such is the multifacetedness of religion. Consequently, I do not adjudicate (i.e., I remain neutral) on religion's possession or lack of these properties. In this way, my analysis uses only a minimal set of assumptions in this regard, and is more charitable towards the degree of scientificity of religion in that I do not deny that religion has any of these properties.

10.4 Chapter Conclusions

Religion lacks at least 66 of the properties of scientificity featured in the demarcatory list, while not clearly having any of the specific ones. This yields a score of −66 for religion. This (possibly provisional) tally is strongly against the scientificity of religion, and is the best explanation for the fact that religion is commonly deemed non-scientific. This, once again, demarcates in a manner that is equivalent to that of the current international philosophical and scientific consensus regarding the extension of the term "science".

References

Bagger, M. C. (2003). Theories of religion. In J. W. V. van Huyssteen (Ed.), *Encyclopedia of science and religion* (2nd ed.). Macmillan Reference.

Clayton, P., & Simpson, Z. (Eds.). (2006). *The Oxford handbook of religion and science*. Oxford University Press.

Coker, R. (2001). *Distinguishing science and pseudoscience*. http://hep.physics.utoronto.ca/~orr/wwwroot/JPH441/Pseudoscience.pdf

De Cruz, H. (2022). Religion and science. In E. N. Zalta & U. Nodelman (Eds.), *The Stanford Encyclopedia of Philosophy* (Fall 2022 ed.). Stanford University. https://plato.stanford.edu/archives/fall2022/entries/religion-science/

Dixon, T. (2008). *Science and religion: A very short introduction*. Oxford University Press.

Ferngren, G. B. (Ed.). (2000). *The history of science and religion in the Western tradition: An encyclopedia*. Routledge.

Glasenapp, H. (1966). *Der Buddhismus – eine atheistische Religion?* Szczesny München.

Kitcher, P. (1982). *Abusing science. The case against creationism*. MIT Press.

Kuhn, T. S. (1977). Objectivity, value judgment, and theory choice. In *The essential tension: Selected studies in scientific tradition and change* (pp. 320–339). The University of Chicago Press.

Lilienfeld, S., Ammirati, R., & David, M. (2012). Distinguishing science from pseudoscience in school psychology: Science and scientific thinking as safeguards against human error. *Journal of School Psychology, 50*(1), 7–36.

Pew Research Center. (2012). *Global religious landscape*. http://www.pewforum.org/2012/12/18/global-religious-landscape-exec/

Pollack, D. (1995). Was ist Religion? Probleme der Definition. *Zeitschrift für Religionswissenschaft, 3*(2), 163–190.

Southgate, C., Deane-Drummond, C., & Murray, P. D. (2011). *God, humanity and the cosmos: A textbook in science and religion* (3rd ed.). T&T Clark/Continuum.

Van Huyssteen, J. W. (Ed.). (2003). *Encyclopedia of science and religion*. Macmillan.

Chapter 11
The Properties of Analytic Philosophy

Abstract This chapter provides a brief definition and a description of/introduction to analytic philosophy. It then meta-analytically shows that the demarcation criteria and the scoring method proposed in Chap. 2 demarcate analytic philosophy as broadly scientific, in agreement with the current philosophical and scientific consensus.

Keywords Analytic philosophy · Scientific demarcation · Philosophy · Science · Scientificity

11.1 Introduction

A certain type of philosophy is seen as scientific in the broad sense (an SBS, or *Wissenschaft*, as understood in Chap. 1) by the authors who favour the concept of SBS (e.g., Hansson, 2013, 2017; Boudry, 2013, 2017; Fernandez-Beanato, 2020; Williamson, 2021).[1] As shown below, analytic philosophy is the best candidate for fulfilling the role of broadly-scientific philosophy. That this is the case can be concluded from the fact that analytic philosophy has historically been "conceived as part of the scientific enterprise", and from the fact that it is currently the only kind of philosophy that is "inside" science. "[T]he analytic tradition is the main home of [broadly] scientific philosophy". (Williamson, personal communication (2021), 2021).

This section provides a working definition of the term "analytic philosophy". Section 11.2 lists properties of science that analytic philosophy has. Section 11.3

[1] "Although the methodological differences between experimental and historical sciences make for fascinating philosophical discussions […], it seems that what distinguishes nuclear physics from cold fusion theory, and Second World War history from Holocaust denial, is an epistemic issue of an altogether different order. The same can be maintained when it comes to philosophy. For example, conceptual equivocation [cf. property 36] is as pernicious in philosophy as it is in science, and the self-protective rationale is exactly the same (Law, 2011)." (Boudry, 2013, p. 93).

D. Fernandez-Beanato, *A Working Scientific Demarcation*, Synthese Library 519, https://doi.org/10.1007/978-3-031-99005-2_11

lists properties of science that analytic philosophy lacks. As the conclusion of this chapter, Sect. 11.4 argues that the demarcation criteria listed in Sect. 2.2 demarcate analytic philosophy as (broadly) scientific, and that the fact that the authors who favour the concept of SBS believe that analytic philosophy is scientific is explained by the fact that it has enough of the properties of science listed in Chap. 2.

The concern of this chapter is what properties of science analytic philosophy has and what properties of science it lacks. This chapter does not provide reasons for believing that analytic philosophy is an SBS other than those that already exist in the literature. I take the properties of science listed in Sect. 2.2 and show that analytic philosophy possesses many of them, according to specialists on analytic philosophy in the context of discussing what analytic philosophy is like. This provides: a) a rational reconstruction (and therefore an explanation) of *why* analytic philosophy is scientific (in the broad sense), b) further evidence of the extensional adequacy of this work, c) further evidence of this work's explanatory power, d) further evidence of the fact that the proposed set of criteria demarcates science appropriately, and e) further evidence of the self-coherence of this work (given that this work claims from its Chap. 1 that certain philosophy is broadly scientific).

This assessment of analytic philosophy uses scholarly literature. This chapter, like all other case studies in this work, incorporates analyses performed by high-quality authors who are specialists in the field and who concluded that analytic philosophy possesses or lacks properties that this work, in turn, uses for SD. I rely here on mainstream authoritative literature on analytic philosophy written by practising philosophers. The texts used are key to this investigation because they engage with the properties of science that are featured in analytic philosophy, *as* they are featured in analytic philosophy. These attributions of properties are mutually commensurable (as part of mainstream traditions in the philosophy of science) and largely uncontroversial. They were arrived at by using different methods and are, in many cases, also independent of our problem, in the sense that the authors' concern when making them was not SD. All of these authors are acknowledged authorities about analytic philosophy, and their works are standard, not controversial, with respect to the points taken up here. For all the above-mentioned reasons, we can rely on this set of authors for our purposes. Therefore, this work recognises that analytic philosophy has the properties that those philosophical authorities adjudicated to it. In this way, this work is meta-analytical, in that its analysis uses and incorporates the conclusions of these authors' original analyses.

Nonetheless, explicit claims or arguments about analytic philosophy's possession of properties of science are difficult to find in the literature, and that is why most of the properties listed in Chap. 2 are not covered in this chapter, although it seems clear that analytic philosophy possesses most of them. In fact, the best examples of analytic philosophy feature nearly all of the applicable properties in the demarcatory list. This is not surprising, since the list is mainly a compilation of proposals made by philosophers of science. Philosophers of science are usually analytic philosophers who deem those properties, were applicable, advisable for all rational endeavours, including both empirical science and philosophy. Therefore, analytic philosophy strives to feature the same demarcatory properties.

Analytic philosophy is the approach to philosophical problems (or the method or "style of practising philosophy" (Haaparanta, 2013)) that "seeks clarity through conceptual analysis" (Bunge, 2003, p. 17). "'[C]onceptual analysis' aims at breaking down complex concepts into their simpler components" (Martinich, 2001, p. 1). The label "analytic philosophy" excludes the frameworks derived from Thomism, as well as "continental philosophy" (i.e., phenomenology, existentialism, and hermeneutics), pragmatism, Hegelianism, Marxism, and postmodernism. "[T]he terms 'analysis,' 'logical analysis,' and 'conceptual analysis' were widely used from the inception of the movement to characterize the methods of philosophy advocated" by analytic philosophers (Hacker, 2002, p. 3). "[T]he methods of analysis and synthesis have a long tradition, starting with ancient geometry. [...] [T]he method of analytic philosophy must be seen in this context." (Haaparanta, 2013). Analytic philosophy must be characterised further, because that of "[']analytical philosophy[' is] too narrow a label, since [it] is not generally [just] a matter of taking a word or concept and analyzing it (whatever exactly that might be)." (McGinn, 2002, p. xi).

Analytic philosophy is a tradition born out of "scientifically-minded philosophers" who tried to purge philosophy of "non-scientific statements" (Preston), which is generally why it is a broadly-scientific CF. It has an "orientation towards science" (Glock, 2013). Let us further characterise analytic philosophy by using quotes from the scholarly literature:

> This tradition emphasizes clarity, rigor, argument, theory, truth. It is not a tradition that aims primarily for inspiration or consolation or ideology. Nor is it particularly concerned with "philosophy of life," though parts of it are. This kind of philosophy is more like science than religion, more like mathematics than poetry—(McGinn, 2002, p. xi)

> [A]nalytic philosophy has especially emphasized the ideal of clarity and pursued exactness by means of logical and mathematical tools [...]. [A]nalytic philosophers think more often than continental philosophers that practising philosophy is like practising science. They think more often than continental philosophers that philosophy has several objects of research that determine the subfields of philosophy and that philosophers use specific research methods, aim at results and construct theories, which the members of the scientific community evaluate. (Haaparanta, 2013)

> Analytic philosophy has been the predominant philosophical movement of the twentieth century. Almost from its inception, it was allied with the spirit of rationality and science, and was dedicated to [...] the eradication of philosophical mystification. Methodologically it was associated with the employment of the new logic as a source of philosophical insight and, somewhat later—after the linguistic turn in philosophy—with a principled and meticulous attention to language and its use. Analytic philosophy flourished in various forms from the 1910s until the 1970s. In the last quarter of the century, however, it has lost its distinctive profile, retaining the name of analytic philosophy largely through its genealogy, the foci of concern which it shares with the antecedent tradition, and its contrastive juxtaposition with certain forms of continental philosophy. [...] [F]rom its inception twentieth-century analytic philosophy differed from its classical seventeenth-century forbears in eschewing psychological analysis and replacing it with logical analysis. The invention of the new logic by Frege, Russell, and Whitehead both set an agenda for analytic philosophy in the first decades of the century and supplied a method. The agenda was to clarify the nature and status of the propositions and laws of logic, to elucidate the relations between Frege's concept-script or Russell's logical language of *Principia* [*Mathematica*] and natural lan-

> guages, and to cast light upon the relation of both natural language and the logical calculus to thought and reality. [... These] questions preoccupied many (but not all) philosophers of the analytic movement. [...] The method (exemplified by Russell's theory of descriptions) consisted in invoking the apparatus of the prepositional and predicate calculi in the endeavor to analyze the subject matter at hand. (Hacker, 2002, pp. 3, 9)

> 'Analytic' philosophy today names a *style* of doing philosophy, not a philosophical program or a set of substantive views. Analytic philosophers, crudely speaking, aim for argumentative clarity and precision; draw freely on the tools of logic; and often identify, professionally and intellectually, more closely with the sciences and mathematics, than with the humanities. (Leiter, 2006)

Let us see how analytic philosophy stands regarding the possession of our scientificity properties.

11.2 The Properties of Analytic Philosophy

Many of the properties that are typical of analytic philosophy were just given in our definition of "analytic philosophy" in Sect. 11.1. The fact that this chapter defines "analytic philosophy" the way it does implies that the chapter is not about any other thing that could conceivably be correctly called "analytic philosophy". Nonetheless, because analytic philosophy has been, and is, a diverse movement, its "readily identifiable" (Hacker, 2002, p. 4) properties "may fail as features uniting *all and only* analytic philosophy, but [...] nevertheless have the merit of characterizing *significant parts* of the analytic movements or important aspects of analytic philosophy." (Glock, 2013). There might be work in analytic philosophy that lacks some of the properties adjudicated in this chapter to analytic philosophy.

It is not necessary that all philosophers who claim or are said to be in the analytic tradition are in fact in it, or that all philosophers who claim to do broadly scientific philosophy in fact do it. It is also not necessary that all analytical philosophers think of what they do as analytic philosophy or as broadly scientific philosophy (even if it is so). And it is not necessary that all philosophers in the analytic tradition always do analytic philosophy or live up to its ideals.

In any case, clearly, there is a substantial body of work (made by a substantial body of philosophers) having properties that correspond to those of the kind of philosophy that is described in this chapter as "analytic". It is because this body of philosophy possesses those properties, that it is deemed broadly scientific by the above-mentioned experts who favour the concept of SBS, and by me. This chapter establishes that insofar as there is this way of doing philosophy, it is broadly scientific (and it is recognised as such), and it is so because it possesses the properties of science listed here. Whether or not this kind of philosophy is a large part of philosophy, or a large part of what is called "analytic philosophy" in other works, or whether or not what is called "analytic philosophy" in other sources has these properties, is not my concern here. This work is also not concerned with what the boundaries of analytic philosophy are. Instead, this work is concerned with the characterisation of the kind of philosophy that is broadly scientific.

Let us list these properties of science possessed by analytic philosophy, which are included in our demarcatory list:

Analytic philosophy emphasises truth (McGinn, 2002, p. xi). Analytic philosophy has "an implicit commitment—albeit faltering and imperfect—to the ideals of clarity, rigor and argumentation", and it "aims at truth and knowledge [...(.)] [T]he goal in analytic philosophy is to discover what is true" (Soames, 2003, pp. xiii–xvii) (properties 3, 69, 194, and 196).

Analytic philosophy features an emphasis on language: "a distinctive feature of much [...] analytic philosophy is its preoccupation with language and linguistic meaning" (Hacker, 2002, p. 12). "It is evident that analytic philosophy has been bound up with a sharpened awareness of the relevance to philosophy of close attention to language and its use." (Hacker, 2002, p. 9). Analytic philosophy is generally meticulous, rigorous, features a high degree of precision and clarity of formulation, and it is painstaking, not vague or ambiguous (properties 87–89, 92, and 94). For example, this work carefully defines many of its key terms, including the names of every field analysed in its case studies, and in some cases needed to untangle the different meanings conveyed by complex polysemous expressions, for example "common sense" (in Chap. 8). This whole work is largely an exercise in explication, definition, clarification, and demarcation.

Analytic philosophy routinely offers accounts (property 42).

"As its name betokens, analytic philosophy is concerned with the analysis of complexes into their constituents" (Hacker, 2002, p. 5). It takes things piecemeal. Analytic philosophy usually uses "the method of offering narrowly focused discussions that probe a specific, isolated issue with precision and attention to detail. [...] [Current] analytic philosophy cannot be defined in terms of a common set of philosophical views or interests, but it can be loosely characterized in terms of its style, which tends to emphasize precision and thoroughness" (Martinich, 2001, p. 1; quote: Preston). Because of the "'rationalist conception' of analytic philosophy which insists that analytic philosophers strive to present reasons for their views" (Føllesdal, 1997; quote: Pincock, 2013), analytic philosophy requires arguments, and places a premium on clarity and rigour in arguments. To this end, analytic philosophy uses formal logic, mathematics, and the natural sciences. It features a "general commitment to analysis [...] and logical construction." (Hacker, 2002, p. 4). It stresses the need for logical and linguistic analysis. It "venerates" (Glock, 2013) formal logic. Analytic philosophy "convinces by appeal to the evidence, [and] by arguments based upon logical and/or mathematical reasoning" (quote: Coker, 2001, p. 6). There is no "disregard for argument" in analytic philosophy. And it recognises the limitations of speculative theorising (Hacker, 2002, p. 3) (properties 51, 65–66, 71, 88–89, 92, 96, 107, 116, 122, and 174).

The sources used for this assessment also make clear that analytic philosophy is a creative activity (property 68).

Analytic philosophy also has a vocation as "critical thinking writ large" (Glock, 2008, p. 260) (property 118).

Regarding the applicability of the properties listed on our list of properties of science, the main difference between empirical science and philosophy is that philosophy is less empirical or less directly empirical, and therefore, the properties that are intimately linked to empirical aspects are applicable to philosophy sometimes only indirectly.

11.3 The Properties That Analytic Philosophy (Mostly) Does Not Have

Analytic philosophy lacks property 191 (as it features a considerable amount of important unsolved problems).

Regarding all the properties in Sect. 2.2 not mentioned in this chapter, it is clear that analytic philosophy mostly has most of them.

11.4 Chapter Conclusions

Analytic philosophy has at least 13 of the specific properties of scientificity featured in the demarcatory list, while lacking at least 1 property. This yields a score of 12 for analytic philosophy. This (provisional) tally is strongly in favour of the scientificity of analytic philosophy, and is the best explanation for the fact that (most) analytic philosophy is commonly deemed scientific by the authors who favour the concept of SBS. This, once again, demarcates in a manner that is equivalent to that of the current international philosophical and scientific consensus regarding the extension of the term "science". This is not a random coincidence, given that, judging from the cited literature, the authors who favour the concept of SBS believe that analytic philosophy is scientific precisely because it has enough of the properties of scientificity listed in Chap. 2.

References

Boudry, M. (2013). Loki's Wager and Laudan's Error. On genuine and territorial demarcation. In M. Pigliucci & M. Boudry (Eds.), *Philosophy of pseudoscience: Reconsidering the demarcation problem* (pp. 79–98). The University of Chicago Press.

Boudry, M. (2017). Plus ultra. Why science does not have limits. In M. Pigliucci & M. Boudry (Eds.), *Science unlimited? The challenges of scientism* (pp. 31–52). University of Chicago Press.

Bunge, M. A. (2003). *Philosophical dictionary (Enlarged edition)*. Prometheus Books.

Coker, R. (2001). *Distinguishing science and pseudoscience*. http://hep.physics.utoronto.ca/~orr/wwwroot/JPH441/Pseudoscience.pdf

Fernandez-Beanato, D. (2020). Cicero's demarcation of science: A report of shared criteria. *Studies in History and Philosophy of Science Part A, 83*, 97–102.

Føllesdal, D. (1997). Analytic philosophy: What is it and why should one engage in it? In H.-J. Glock (Ed.), *The rise of analytic philosophy*. Blackwell.

Glock, H.-J. (2008). *What is analytic philosophy?* Cambridge University Press.

Glock, H.-J. (2013). Replies to my Commentators. *Journal for the History of Analytical Philosophy, 2*(2).

Haaparanta, L. (2013). Philosophy and its Recent History: Remarks on What is Analytic Philosophy?, by Hans-Johann Glock. *Journal for the History of Analytical Philosophy, 2*(2).

Hacker, P. M. S. (2002). Analytic philosophy: What, whence, and whither? In A. Biletzki & A. Matar (Eds.), *The story of analytic philosophy: Plot and heroes*. Routledge.

Hansson, S. O. (2013). Defining pseudoscience and science. In M. Pigliucci & M. Boudry (Eds.), *Philosophy of pseudoscience: Reconsidering the demarcation problem* (pp. 61–78). The University of Chicago Press.

Hansson, S. O. (2017). Science and pseudo-science. In E. N. Zalta (Ed.), *The Stanford Encyclopedia of Philosophy* (Summer 2017 ed.). Stanford University. https://plato.stanford.edu/archives/sum2017/entries/pseudo-science/

Law, S. (2011). *Believing bullshit: How not to get sucked into an intellectual black hole*. Prometheus Books.

Leiter, B. (2006). *"Analytic" and "Continental" philosophy*. https://web.archive.org/web/20061115002425/https://www.philosophicalgourmet.com/analytic.asp

Martinich, A. P. (2001). Introduction. In A. P. Martinich & D. Sosa (Eds.), *A companion to analytic philosophy*. Blackwell.

McGinn, C. (2002). *The making of a philosopher: My journey through Twentieth-century philosophy*. HarperCollins.

Pincock, C. (2013). On Hans-Johann Glock, What is analytic philosophy? *Journal for the History of Analytical Philosophy, 2*(2).

Preston, A. Analytic philosophy. In *Internet Encyclopedia of Philosophy*. https://iep.utm.edu/analytic/

Soames, S. (2003). *The dawn of analysis* (2nd print., 1st papers. print. ed.). Princeton University Press.

Williamson, T. (2021). *Degrees of freedom: Is good philosophy bad science?*. Lecture read to the LanCog Research Group, Centro de Filosofia, Faculdade de Letras da Universidade de Lisboa, on 7th October, 2021.

Chapter 12
The Properties of History

Abstract This chapter provides a brief definition and a description of/introduction to the discipline of history. It then meta-analytically shows that the demarcation criteria and the scoring method proposed in Chap. 2 demarcate history as broadly scientific, in agreement with the current philosophical and scientific consensus.

Keywords History · Scientific demarcation · Science · Humanities · Scientificity

12.1 Introduction

Most of the current mainstream discipline (CF) of human history[1] (referred to in this work simply by the term "history") is scientific in the broad sense (Bloch, 1992 [1949]; Carr, 1961; Ajaegbo, 2013; Mahner, 2013, p. 35; Hansson, 2013, p. 64; Boudry, 2013, pp. 92–4[2]). This chapter shows that the demarcation criteria presented in Chap. 2 demarcate history as scientific, in agreement with that consensus. This section provides a working definition of "history". Section 12.2 lists properties of scientificity that history has. Section 12.3 lists properties of scientificity that

[1] I specify "human history" because there are other historical sciences, e.g., palaeontology. Bloch put it this way: "It is sometimes said: 'History is the science of the past'. To me, this is badly put. […] [T]here is a history of the solar system, because the [… (celestial bodies)] which compose it have not always been as we now see them. It belongs to the province of astronomy. There is a history of volcanic eruptions which is, I am sure, of most lively interest as regards the composition of the earth. It does not concern the history of historians. Or, at least, it does so only in so far as its observations chance to coincide with the specific preoccupations of our history. […] What is it that seems to dictate the intervention of history? It is the appearance of the human element." (Bloch, 1992 [1949], pp. 19–21).

[2] "Although the methodological differences between experimental and historical sciences make for fascinating philosophical discussions […], it seems that what distinguishes nuclear physics from cold fusion theory, and Second World War history from Holocaust denial, is an epistemic issue of an altogether different order. […] For example, conceptual equivocation is as pernicious in philosophy as it is in science, and the self-protective rationale is exactly the same (Law, 2011)." (Boudry, 2013, p. 93).

D. Fernandez-Beanato, *A Working Scientific Demarcation*, Synthese Library 519, https://doi.org/10.1007/978-3-031-99005-2_12

history lacks. As the conclusion of this chapter, Sect. 12.4 argues that the demarcation criteria listed in Sect. 2.2 demarcate history as (broadly) scientific, and that the fact that the authors who favour the concept of SBS believe that history is scientific is explained by the fact that it has enough of the properties of science listed in Chap. 2.

The concern of this chapter is what properties history has and what properties it lacks. This chapter does not provide reasons for believing that history is an SBS other than those that already exist in the literature. I take the properties of science listed in Sect. 2.2 and show that history possesses many of them, according to specialists on history in the context of discussing what history is like. This provides: a) a rational reconstruction (and therefore an explanation) of *why* history is scientific (in the broad sense), b) further evidence of the extensional adequacy of this work, c) further evidence of this work's explanatory power, d) further evidence of the fact that the proposed set of criteria demarcates science appropriately, and e) further evidence of the self-coherence of this work (given that this work claims from its Chap. 1 that certain history is broadly scientific).

This assessment of history uses scholarly literature. This chapter, like all other case studies in this work, incorporates analyses performed by high-quality authors who are specialists in the field and who concluded that history possesses or lacks properties that this work, in turn, uses for SD. I rely here on mainstream authoritative literature on history written by practising historians. The texts used are key to this investigation because they engage with the properties of science that are featured in history, *as* they are featured in history. These attributions of properties are mutually commensurable (as part of mainstream traditions) and largely uncontroversial. They were arrived at by using different methods and are, in many cases, also independent of our problem, in the sense that the authors' concern when making them was not SD. These authors are acknowledged authorities about history, and all of their works are standard, not controversial, with respect to the points taken up here. Also, these authors' works are rooted in detailed engagement with academic human history, which adds to their reliability. For example, Bloch is one of the most celebrated practising historians of the Twentieth century. Consequently, this chapter draws especially upon his *The Historian's Craft*. For all the above-mentioned reasons, we can rely on this set of authors for our purposes. Therefore, this work recognises that history has the properties that those authorities adjudicated to it. In this way, this work is meta-analytical, in that its analysis uses and incorporates the conclusions of these authors' original analyses.

Nonetheless, explicit claims or arguments about history's possession of properties of science are difficult to find in the literature, and that is why most of the properties listed in Chap. 2 will not be covered in this chapter, although it seems clear that history possesses most of them. In fact, the best examples of history feature nearly all of the applicable properties listed in the demarcatory list.

History is the (broadly) scientific CF which is about the human past (the same expression can be used to refer to that past itself, but we are not concerned with it here). It encompasses the discovery, collection, organisation, presentation, and interpretation of information about human past events. It describes, examines, questions, and analyses past events related to humans, including their consequences, and investigates their causes and effects.

More narrow definitions of "human history" conventionally make its period of study start during the fifth or fourth millennium BCE (depending on which civilisational milestone is chosen, the most common being the formation of the first complex societies or the first states, the undertaking of the first large-scale public works, or the first invention of writing, all of which occurred roughly in this period).

When referring to "human history", this work ignores this difference between a history of all the human past and a history of only its latest 6500 years or so, because such conventional difference has no substantial bearing on the degree of scientificity of the CF (although it does bear strongly on the availability of different types of sources, or lack thereof), nor on the assessment of that degree, with which we are concerned here.

The fact that this chapter defines "history" the way it does implies that the chapter is not about any other thing that could conceivably be called "history". There might be work in history that lacks some of the properties adjudicated in this chapter to history. And it is not necessary that all historians who claim or are said to be broadly scientific historians are in fact broadly scientific historians, or that all historians who claim to do broadly scientific history in fact do it. It is also not necessary that all broadly scientific historians think of what they do as broadly scientific (even if it is so). And it is not necessary that all broadly scientific historians always do broadly scientific history or live up to its ideals.

The authoritative historians referenced in this chapter agree with me that all of history that is worth the name is broadly scientific. But even without committing to this position we can clearly see that a large body of the work of history (made by a substantial body of historians) is broadly scientific (in line with the definitions given in Chap. 1), and recognise that a large body of historians say so. This body of work is of the kind of history that this chapter is concerned with. It is deemed broadly scientific by the above-mentioned experts (who favour the concept of SBS) and by me because it possesses the properties of science listed in this chapter, i.e., the properties of the kind of history that is described in this chapter as broadly scientific. This chapter argues that insofar as there is this way of doing history, it is broadly scientific (and it is recognised as such), and it is so because it possesses the properties of science listed here. Whether or not this kind of history is a large part of all history, or a large part of what is called "history" in other works, or whether or not what is called "history" in other sources has these properties, is not my concern here. Instead, this work is concerned with the characterisation of the kind of history that is broadly scientific.

Let us see how history stands regarding the possession of our scientificity properties.

12.2 The Properties of History

If the historian sufficiently "employ[s] the canons of historical scholarship" (Ajaegbo, 2013, p. 12), then the history she presents is objective (property 2): historical work is based on "objective empiricism" (ibid., p. 10).

History generates (and has generated) hypotheses and theories through deductive and non-deductive processes to try to get to the truth (properties 3 and 31):

> [I]n the pursuit of his vocation, the historian draws from many primary sources, employs the knowledge of other disciplines and tries to be as scientific or empirical as possible in his quest to establish historical truth. […] The historian also uses both inductive and deductive methods to explain how and why human beings took certain actions in the past and the consequences of such actions. (Smitha, 2005, p. 2; quote: Ajaegbo, 2013, p. 1)

> Certainly, we would or should want to preserve the idea of objective knowledge for historical study. A historian with postmodernist or poststructuralist sympathies like Hayden White might not find it a cause for concern, but for most historians the survival of the discipline would be at stake if, without a belief in objectivity, it were to become simply a species of fiction. According to Appleby, Hunt and Jacob [(1994)], "history is crucially distinguished from fiction by curiosity about what actually happened in the past" ([…] p. 259). The crucial word here is "actually," [...] What distinguishes history are the constraints it imposes upon itself to accumulate and not violate the facts, even as it goes beyond them in interpretation and speculation. (Goodheart, 2005, p. 486)

History is useful for harbouring conceptions that correspond to facts (property 50). It offers accounts that have power to explain intended data and that are based on its collection of pieces of data (Bloch, 1992 [1949], pp. 43, 59, 62; Smitha, 2005; Ajaegbo, 2013, pp. 8, 12[3]). "[H]istorical methods are systematic, sequential, logical and progress in clearly defined steps" (ibid., p. 1). History analyses (Bloch, 1992 [1949], Chapter 4; Ajaegbo, 2013, p. 8), classifies, generalises, synthesises, and it is holistic, treating reality systemically (properties 42, 95, 107, and 116):

> [T]he true function of analysis in any category of research [(…) is to dissect] reality only in order to observe it better […]. Fustel de Coulanges […] announced to his listeners in the Sorbonne: "Supposing a hundred specialists had divided the past of France according to lot, do you think that, in the end, they would have written the history of France? I very much doubt it. At the very least, they should miss the linkage of facts: now, this linkage is itself a historical truth." […] [A] civilization, like a person, is no mechanically arranged game of solitaire; the knowledge of fragments, studied by turns, each for its own sake, will never produce the knowledge of the whole; it will not even produce that of the fragments themselves.
>
> But the work of reintegration can come only after analysis. Better still, it is only the continuation of analysis, and its ultimate justification. In the scene as we first perceive it,

[3]"The historian collects his facts from oral traditions, songs, inscriptions, documentary sources and artifacts which are often kept in museums." (Smitha, 2005, p. 2; quote: Ajaegbo, 2013, p. 1). "A historian collects evidence or historical facts for the study of the past. Historical facts are available to the historian in oral forms, documents, artifacts or material remains of man. Evidence constitutes the raw materials of history. The historian collects his facts by asking questions of how, why, what, when and where." (Ajaegbo, 2013, p. 8).

which we contemplate rather than observe, and in which therefore nothing is distinct, we cannot discern the interrelationships. Their delicate network becomes visible only after the elements have been classified. Moreover, in order to remain true to life in its intertwining actions and reactions, we need not pretend to seize it as a whole. Such an attempt is too vast for the powers of a single scholar. Nothing is more legitimate, nothing is more salutary than to center the study of a society upon one of its particular aspects, or, better still, upon one of the well-defined problems underlying one or another of these aspects: the beliefs, the economy, the class or group structure, the political crises....

By this systematic selection, not only will the problems usually be more concretely stated, but even the facts of connection and interchange will stand out with greater clarity. (Bloch, 1992 [1949], pp. 124, 128–9)

History is generally fallible/provisional/tentative (property 63): "Historical and scientific knowledge or conclusions are not absolute. Historical and scientific ["]facts["] undergo revision or rejection, in the light of new evidence." (Ajaegbo, 2013, p. 11).

History is a creative activity (property 68): "The historian [...] seek[s] to discover what is not yet known. [(...) She is a] creative thinker" (ibid., p. 16).

History features fecundity/fruitfulness, progress (historical knowledge is "subject to continual development and refinement" (Marwick, 1989, p. 21)), and promise (properties 82–84):

[T]he knowledge of the past is something progressive which is constantly transforming and perfecting itself. Anyone who doubts this need only recall how much it has improved under our very eyes in little more than a century. Vast areas of mankind have emerged from the shadows. Egypt and Chaldea have shaken off their shrouds. The lost cities of central Asia have disclosed their now-unspoken languages and long-extinct religions. A civilization, all unsuspected, has but lately risen from its grave upon the banks of the Indus. That is not all, and the ingenuity of the scholars in further ransacking the libraries or in opening new excavations on ancient sites is neither the sole nor, perhaps, even the most effective means of enriching our picture of the past. Hitherto-unknown techniques of investigation have also come to light. We are more skillful than our predecessors in examining languages for the evidence of customs and tools for the evidence of techniques. Above all, we have learned how to probe more deeply in the analysis of social developments. The study of popular rites and beliefs is barely sketching its first outlines. [...] All this is certain. All of it offers us the most extensive hopes [...]. (Bloch, 1992 [1949], pp. 48–9)

History's method is rigorous and precise (properties 88–89): "the historian, no less than the physicist, employs precise and rigorous methods in his enquiries" (Haddock, 1980, p. 151). History uses "systematic methods [...] involving rigorous checks and the presentation of evidence as well as conclusions." (Marwick, 1989, p. 152) (properties 98 and 102).

History is painstaking (Bloch, 1992 [1949], p. 161) and features screening and sorting of putative facts that is appropriate to separate pseudofacts from facts (ibid., pp. 75–91) (property 99): "Like any scholar, [...] the historian selects and sorts." (ibid., p. 119).

"History is [...] scientific in its method and technique" (Ajaegbo, 2013, p. 16) (properties 51 and 103–104). This includes the fact that history routinely follows an hypothetico-inferential method, making inferences on the bases of data and hypotheses. For example, if a historian believes that a big battle was fought between the

Anglo-Saxons and the Norman-French 11 km to the northwest of Hastings in 1066, then she can predict that, if a proper search is conducted, the remains of weaponry featuring certain particular properties will probably be found buried there (otherwise, historians would not know what to look for, they would never know where to excavate, or what archives to look in).

History is critical and employs critical thinking (Haddock, 1980, pp. 3–4; Bloch, 1992 [1949], Chapter 3; Ajaegbo, 2013, p. 10).

> Facts are of many orders, and what is crucial to historical study is the nature of evidence. [...] This evidence is to be found in the sources, relics and traces left by the past. The fundamental historical skills lie in the analysis, criticism, and interpretation of these sources, which exist in an enormous variety [...] (Marwick, 1989, p. 380)

"The scholar records—better still, he invites—the experience which may, perhaps, upset his most cherished theories." (Bloch, 1992 [1949], p. 114). "Only through critical methods and inter-disciplinary approach to historical studies can a fairly accurate, verifiable and acceptable historical knowledge be produced." (Ajaegbo, 2013, p. 12). History "use[s] critical methods to [...] re-construct the past of [hu]mankind" (ibid., p. 16) (properties 51, 118–119, and 122).

History is connected and contiguous with other scientific fields: "[I]n the pursuit of his vocation, the historian [...] employs the knowledge of other disciplines" (Smitha, 2005, p. 2; quote: Ajaegbo, 2013, p. 1) (properties 59, 109, and 189). For example, the composition of a metallic coin can help historians establish its origins, genetics can illuminate the magnitude of historical population exchanges, and astronomy can date a document mentioning the apparition of a nova.

History looks for the causes (Bloch, 1992 [1949], p. 7 and Chapter 5; Goodheart, 2005, p. 483) and the composition, structure, and the kind and intensity of effects of things (these factors can then be theorised about more abstractly in sociological theory (covered in Chap. 19)) (property 197):

> Whether he likes it or not, every physicist, every biologist thinks in terms of "why" and "because." The historian cannot escape this common law of the mind. [...] [T]he genetic link is present throughout. [... The] establishment of relations of cause and effect constitutes an instinctive need of our understanding [...] [T]he use of the causal relationship [... (is)] a tool of historical knowledge. (Bloch, 1992 [1949], p. 157)

Regarding the applicability of the properties listed on our list of properties of science, history cannot conduct experiments, and cannot directly observe the phenomena it studies (except for history of the present), so history is less directly empirical than other empirical SBSs, and therefore, the properties that are intimately linked to empirical aspects are applicable to history sometimes only indirectly.

12.3 The Properties That History (Mostly) Does Not Have

Because history does not generate general laws (although sociology might), it lacks the properties 181–182 and 195.

Some authors argue that history is less successful at predicting the future than other sciences, but whether it is so is not clear enough to warrant tallying properties related to prediction against the degree of scientificity of history. Historians, having more reliable data at their disposal, make predictions about the future that are more epistemically reliable and, on average, more successful than the predictions of non-scientific CFs such as common sense and pseudoscience. And other sciences are not as successful at predicting as it is sometimes implied; e.g., physics cannot predict where a banknote dropped from the top of the Empire State Building will land.

Regarding all the properties in Sect. 2.2 not mentioned in this chapter, it is clear that history mostly has most of them. Explicit claims or arguments in this regard are difficult to find in the literature precisely because the possession of these properties by history is usually taken for granted.

12.4 Chapter Conclusions

History has at least 16 of the specific properties of scientificity featured in the demarcatory list, while lacking at least 3 properties. This yields a score of 13 for history. This (possibly provisional) tally stands strongly in favour of the scientificity of history, and is the best explanation for the fact that (most) history is commonly deemed scientific by the authors who favour the concept of SBS. This, once again, demarcates in a manner that is equivalent to that of the current international philosophical and scientific consensus regarding the extension of the term "science". This is not a random coincidence, given that, judging from the cited literature, the authors who favour the concept of SBS believe that history is scientific precisely because it has enough of the properties of scientificity listed in Chap. 2.

References

Ajaegbo, D. I. (2013). The unity of knowledge: History as science and art. *AFRREV IJAH: An International Journal of Arts and Humanities, 2*(3).

Appleby, J., Hunt, L., & Jacob, M. (1994). *Telling the truth about history*. W. W. Norton.

Bloch, M. (1992 [1949]). *The historian's craft*. Manchester University Press.

Boudry, M. (2013). Loki's Wager and Laudan's Error. On genuine and territorial demarcation. In M. Pigliucci & M. Boudry (Eds.), *Philosophy of pseudoscience: Reconsidering the demarcation problem* (pp. 79–98). The University of Chicago Press.

Carr, E. H. (1961). *What is history?* Penguin Books.

Goodheart, E. (2005). Is history a science? *Philosophy and Literature, 29*(2), 477–488.

Haddock, B. A. (1980). *An introduction to historical thought*. Edward Arnold.

Hansson, S. O. (2013). Defining pseudoscience and science. In M. Pigliucci & M. Boudry (Eds.), *Philosophy of pseudoscience: Reconsidering the demarcation problem* (pp. 61–78). The University of Chicago Press.

Law, S. (2011). *Believing bullshit: How not to get sucked into an intellectual black hole*. Prometheus Books.

Mahner, M. (2013). Science and pseudoscience. How to demarcate after the (alleged) demise of the demarcation problem. In M. Pigliucci & M. Boudry (Eds.), *Philosophy of pseudoscience: Reconsidering the demarcation problem* (pp. 29–44). The University of Chicago Press.
Marwick, A. (1989). *The nature of history* (3rd ed.). Macmillan.
Smitha, F. E. (2005). *Good history, science or art?* www.macrohistory.org/com/sci-art.htm

Chapter 13
The Properties of the Formal Sciences

Abstract This chapter provides a brief definition and a description of/introduction to the formal sciences. It then meta-analytically shows that the demarcation criteria and the scoring method proposed in Chap. 2 demarcate the formal sciences as scientific, in agreement with the current philosophical and scientific consensus.

Keywords Formal sciences · Scientific demarcation · Science · Logic · Mathematics · Scientificity

13.1 Introduction

This section provides a working definition of the term "formal sciences". Section 13.2 lists properties of science that the formal sciences have. Section 13.3 lists properties of science that the formal sciences lack. As the conclusion of this chapter, Sect. 13.4 argues that the demarcation criteria listed in Sect. 2.2 demarcate the formal sciences as (broadly) scientific, and that the fact that the authors who favour the concept of SBS believe that the formal sciences are scientific is explained by the fact that they have enough of the properties of science listed in Chap. 2.

The formal sciences are the SBSs that are only about abstract structures (as opposed to the natural sciences, psychology, the social sciences, and their multidisciplinary combinations, which are about particular empirical realities). Some examples of formal sciences are: logic (including classical logic, modal logic, parts of philosophical logic, etc.), mathematics (as classically exemplified by arithmetic, geometry, and analysis (Bunge, 1985, p. 12), and including probability theory), and other fields, including statistics, systems science, computer science, artificial intelligence studies, information theory, game theory, decision theory, and theoretical linguistics, or parts of them (Löwe, 2002, p. 6).

Many basic philosophical contentions about the formal sciences are left aside in this chapter (e.g., whether there is synthetic a priori knowledge, whether the formal sciences are a priori or based empirically, whether they give us knowledge of the world, whether they are a problem for empiricism, whether their conclusions can be

D. Fernandez-Beanato, *A Working Scientific Demarcation*, Synthese Library 519, https://doi.org/10.1007/978-3-031-99005-2_13

revised, as Quine thought; the debate about the degree of exceptionality, if any, or the formal sciences (or, specifically, of logic); the debate logical pluralism–logical monism). In any case, the answers to those matters do not bear on the matters discussed in this chapter.

Everyone agrees that the formal sciences are sciences in the broad sense (SBSs), which is why they are called "formal *sciences*" (German: *Formalwissenschaft*). In the case of their name in English, the component term "science" is used literally, although not to mean "science" in a narrow sense (e.g., "natural science"), but to mean "SBS" (i.e., *Wissenschaft*). For example, mathematics is sometimes known as "the queen of the sciences", and there are departments of Mathematical Science in universities. Mathematics overlaps with the natural sciences in disciplines such as mathematical physics and mathematical biology.

The concern of this chapter is what properties of science the formal sciences have and what properties of science they lack. This chapter does not provide reasons for believing that the formal sciences are SBSs other than those that already exist in the literature. I take the properties of science listed in Sect. 2.2 (except those that are intimately linked to empirical aspects, e.g., experimentation[1]) and show that the formal sciences possess many of them, according to specialists on the formal sciences in the context of discussing what the formal sciences are like. This provides: a) a rational reconstruction (and therefore an explanation) of *why* the formal sciences are scientific (in the broad sense), b) further evidence of the extensional adequacy of this work, c) further evidence of this work's explanatory power, and d) further evidence of the fact that the proposed set of criteria demarcates science appropriately.

This assessment of the formal sciences uses scholarly literature. This chapter, like all other case studies in this work, incorporates analyses performed by high-quality authors who concluded that the formal sciences possess or lack properties that this work, in turn, uses for SD. I rely here on mainstream authoritative literature on the formal sciences. The texts used are key to this investigation because they engage with the properties of science that are featured in the formal sciences, *as* they are featured in the formal sciences. These attributions of properties are mutually commensurable (as part of mainstream traditions in the philosophy of science) and largely uncontroversial. They were arrived at by using different methods. All of these authors are acknowledged authorities about the formal sciences, and their works are standard, not controversial, with respect to the points taken up here. For example, Bunge was one of the most relevant philosophers who addressed problems

[1] The main difference between the formal sciences and empirical SBSs is that the formal sciences are not empirical (they do not even claim to advance claims about empirical matters), or are less directly empirical (more abstract) than the empirical SBSs. Because of that, properties that are intimately linked to empirical aspects are not applicable, or are applicable only indirectly, to the formal sciences. In this work those properties are considered inapplicable to the formal sciences, and therefore the lack of those properties is not computed against the scientificity of the formal sciences. But in any case, even if one does count the lack of these properties against the scientificity of the formal sciences, then in the final tally the formal sciences still make the cut as sciences according to our scoring method.

of SD, including the demarcation of the formal sciences. Consequently, this chapter draws especially upon his *Treatise on Basic Philosophy*. For all the above-mentioned reasons, we can rely on this set of authors for our purposes. Therefore, this work recognises that the formal sciences have the properties that those authorities adjudicated to them. In this way, this work is meta-analytical, in that its analysis uses and incorporates the conclusions of these authors' original analyses.

Nonetheless, explicit claims or arguments about the possession of properties of science by the formal sciences are difficult to find in the literature because the possession of these properties by the formal sciences is usually taken for granted, and that is why most of the properties listed in Chap. 2 will not be covered in this chapter, although it seems clear that the formal sciences possess most of them. In fact, the best examples of the formal sciences feature nearly all of the applicable properties in the demarcatory list.

Let us see how the formal sciences stand regarding the possession of our scientificity properties.

13.2 The Properties of the Formal Sciences

The formal sciences possess properties 16, 37–38, and 202–203:

> Whatever the explicit philosophy (if any) that a mathematician may espouse, he works on a *tacit philosophy* including the following tenets. Firstly, he treats the objects of his research (functions, spaces, or what have you) as *sui generis* objects devoid of physical properties. In fact he proceeds *as if* they were self-existent in a world of their own, to the point that he prefers to say that he discovers or finds his results rather than claiming that he has constructed or invented them. Secondly, his work is theoretical: he deals with constructs, not concrete things; consequently he makes neither measurements nor experiments. Thirdly, he abides—just like the factual scientist—by a moral code that commands not to make up anything arbitrarily (except perhaps definitions), not to plagiarize, and not to keep the results of his work from others. (Bunge, 1985, p. 14)

The formal sciences are systematic and analytical (Bunge, 1985, p. 72). They are generally meticulous, rigorous ("i.e.[,] precise and consistent" (ibid.)), feature a high degree of precision and clarity of formulation, and are painstaking, not vague or ambiguous. They place a premium on clarity and rigour. They feature a general commitment to logical construction. They convince "by appeal to the evidence, [and] by arguments based upon logical and/or mathematical reasoning" (Coker, 2001, p. 6). They do not shun conceptually and logico-mathematically exact thinking. The formal sciences have a formal background; "the formal background [...] of mathematics [...] is, trivially, mathematics itself." (Bunge, 1985, p. 14) (properties 29, 35–36, 39, 65–66, 88, 92, 107, 116, 174, 178, 193, and 195).

Property 41:

> The *fund of knowledge* [...] of mathematics [...] is the collection of results obtained previously by mathematicians and which are still valid. As is well known, the rate of decay or attrition of mathematical knowledge is the slowest. True, modern mathematicians have

> been able to improve on the work of Apollonius, Eudoxus, Archimedes, Euclid, and other ancient mathematicians. Still, that work stands on the whole; this cannot be said of ancient physics, with the sole exception of fragments of statics, hydrostatics, and ray optics. (Bunge, 1985, p. 15)

The same can be said of logic: e.g., the bulk of Aristotelian logic is still valid.

All empirical scientific CFs use the formal sciences. Properties 59 and 189:

> The *continuity* condition [...] is satisfied both locally, i.e. by every branch of mathematics, and globally, by mathematics as a whole. In fact there are no isolated chapters of mathematics: the latter makes up a conceptual supersystem. Consequently every mathematical specialist can talk to colleagues in at least one other field of mathematics. Moreover, mathematics interacts vigorously with factual science and technology: it lends them formal tools, and receives in exchange new problems and new tools—namely calculators and computers. Mathematics is thus at the very center of the science and technology system: though conceptually self-contained, it is not self-sufficient, let alone selfish. This is why those who speak the language of mathematics can talk to specialists in all the advanced sciences and technologies. Only philosophers can compete with mathematicians in universality—particularly if they understand the universal language. (Bunge, 1985, p. 16)

Properties 83–84:

> The greatest philosophical interest in the formal sciences is surely the promise they hold of necessary, provable knowledge which is at the same time about the real world [at least in their applied forms], not just some Platonic or abstract idealisation of it. (Franklin, 1994, p. 27)

Properties 103–104:

> The *methodics* [(...) of the formal sciences] is composed, like that of [(…) empirical] science, of the general method of science plus an expanding collection of special methods or techniques. The former boils down to the sequence: *Formulate the problem-Invent hypotheses or theories-Test the solution*. The last step, testing, may consist in constructing a proof, finding a counter-example, or merely checking whether the candidate (e.g. the presumed theorem) satisfies the conditions of the problem. All such operations are exclusively conceptual, even though they may require the assistance of instruments other than the brain [...] (Bunge, 1985, p. 15)

The formal sciences use proofs (property 154).

Property 176, about change as a result of research results obtained in and through scientific inquiry, is possessed by the formal sciences "as long as there are active research [(…) specialists], because the job of every investigator is to innovate in his field, not just to teach it or use it." (Bunge, 1985, p. 16).

Property 200:

> Research [in the formal sciences] is done by members of a mathematical community hosted by a sufficiently advanced society. The community is held together by a tradition and by information flows. [...] [R]esearch [in the formal sciences] (as distinct from the invention of a few crude and stray mathematical ideas) did not start before the dawn of civilization [...] (Bunge, 1985, p. 14)

13.3 The Properties That The Formal Sciences (Mostly) Do Not Have

No applicable property of science seems to be clearly lacking in the formal sciences.

Regarding all the properties in Sect. 2.2 not mentioned in this chapter, it is clear that the formal sciences mostly have most of the applicable ones.

13.4 Chapter Conclusions

The formal sciences have at least 16 of the specific properties of scientificity featured in the demarcatory list, while not clearly lacking any applicable property. This yields a score of 16 for the formal sciences. This (provisional) tally is strongly in favour of the scientificity of the formal sciences, and is the best explanation for the fact that the formal sciences are commonly deemed scientific. This, once again, demarcates in a manner that is equivalent to that of the current international philosophical and scientific consensus regarding the extension of the term "science". This is not a random coincidence, given that, judging from the cited literature, the authors who favour the concept of SBS believe that the formal sciences are scientific precisely because they have enough of the properties of scientificity listed in Chap. 2.

References

Bunge, M. A. (1985). *Treatise on basic philosophy, Vol. 7: Epistemology and methodology III: Philosophy of science and technology*. D. Reidel.

Coker, R. (2001). *Distinguishing science and pseudoscience*. http://hep.physics.utoronto.ca/~orr/wwwroot/JPH441/Pseudoscience.pdf

Franklin, J. (1994). The formal sciences discover the philosophers' stone. *Studies in History and Philosophy of Science, 25*(4), 513–533.

Löwe, B. (2002). The formal sciences: Their scope, their foundations, and their unity. *Synthese, Foundations of the Formal Sciences I, 133*(1/2), 5–11.

Chapter 14
The Properties of Physics

Abstract This chapter provides a brief definition and a description of/introduction to physics. It then meta-analytically shows that the demarcation criteria and the scoring method proposed in Chap. 2 demarcate physics as scientific, in agreement with the current philosophical and scientific consensus.

Keywords Physics · Scientific demarcation · Science · Logic · Mathematics · Scientificity

14.1 Introduction

Virtually everyone believes that current academic physics is scientific. It is an observational and mature natural science (Young & Stadler, 2018, p. 1) that was considered paradigmatically (in the everyday, not the Kuhnian sense of this word) scientific by classical philosophers of science, such as Carnap, Hempel, Popper ("the most complete realization to date of what I call 'empirical science'" (1959/2002, p. 15)), and Lakatos, up to and including more sceptical philosophers, such as Kuhn. This section provides a working definition of "current academic physics". Section 14.2 lists properties of science that it has. Section 14.3 lists properties of science that it lacks. As the conclusion of this chapter, Sect. 14.4 argues that the demarcation criteria listed in Sect. 2.2 demarcate current academic physics as scientific, and that the fact that virtually everyone believes that current academic physics is scientific is explained by the fact that it has enough of the properties of science listed in Chap. 2.

Current academic physics is a CF that studies matter and radiation, and their behaviour in space and time, using concepts such as "energy" and "force". Regarding its objects of study, "[p]hysics […] is the most fundamental of the sciences." (Young & Stadler, 2018, p. 1; Griffith & Brosing, 2022, p. 7).

As with all scientific CFs, a physics is scientific as long as it possesses enough of the properties of scientificity and is the one that contextually (in a given period and society) has the most of them (in this way, previous developments stop being

D. Fernandez-Beanato, *A Working Scientific Demarcation*, Synthese Library 519, https://doi.org/10.1007/978-3-031-99005-2_14

scientific as soon as a (demonstrably) sufficiently epistemically-progressive integral separated approach develops).

Physics began in Antiquity, was a part of the scientific revolution that started in the Sixteenth century, and, after the triumph of Newtonianism, suffered a discipline-specific further revolution at the beginning of the Twentieth century, when the conviction in absolute space and time and classical mechanics was replaced by relativity theory and quantum mechanics.

From now on, this chapter will use the term "physics" to refer to current academic physics.

The concern of this chapter is what properties of science physics has and what properties of science it lacks. This chapter does not provide reasons for believing that physics is a science other than those that already exist in the literature. I take the properties of science listed in Sect. 2.2, show that physics possesses many of them, and that this is asserted by the specialists on physics in the context of discussing what physics is like. This provides: a) a rational reconstruction (and therefore an explanation) of *why* physics is scientific, b) further evidence of the extensional adequacy of this work, c) further evidence of this work's explanatory power, and d) further evidence of the fact that the proposed set of criteria demarcates science appropriately.

This work assesses physics by using scholarly literature. This chapter, like all other case studies in this work, incorporates analyses performed by high-quality authors who concluded that physics possesses or lacks properties that this work, in turn, uses for SD. I rely here on mainstream authoritative literature on physics, including introductory textbooks. Some of the texts used here are issued by prestigious international academic publishers, are extensively utilised in higher education, have had numerous editions, and are of the latest textbooks in the field (e.g., Griffith & Brosing, 2022; Young & Stadler, 2018). The texts used are key to this investigation because they engage with the properties of science that are featured in physics, *as* they are featured in physics. These attributions of properties are mutually commensurable (as part of mainstream traditions in the philosophy of science) and largely uncontroversial. They were arrived at by using different methods and are, in many cases, also independent of our problem, in the sense that the authors' concern when making them was not SD. All of these works are sufficiently up-to-date, and are standard, not controversial, with respect to the points taken up here. For all these reasons, we can rely on this set of authors for our purposes. Therefore, this work recognises that physics has the properties that those scientific authorities adjudicated to it. In this way, this work is meta-analytical, in that its analysis uses and incorporates the conclusions of these authors' original analyses.

Nonetheless, explicit claims or arguments about physics' possession of properties of science are difficult to find in the literature because the possession of these properties by physics is usually taken for granted, and that is why most of the properties listed in Chap. 2 will not be covered in this chapter, although it seems clear that physics possesses most of them.

Let us see how physics stands regarding the possession of our scientificity properties.

14.2 The Properties of Physics

Physics' propositions are confirmable and disconfirmable, and it features confirmations and disconfirmations (e.g., Griffith & Brosing, 2022, pp. 5, 62, 74, 185, 187, 224, 310, 317–8, 333, 417, 447–8, 451–2, 460) (properties 7, 51–52, and 54).

Physics generalises (properties 23, 32, and 107) (Griffith & Brosing, 2022, pp. 4–6). It generates or has generated hypotheses and theories through deductive and non-deductive processes (Grodzicki, 2021, p. 49) (property 31). It offers accounts, describes (ibid., pp. 50, 52, 121–35), and has the ability to be used to causally explain, having broad power to explain observable phenomena (Margenau, 1935, pp. 64–70; Popper, 1959/2002, p. 39; Griffith & Brosing, 2022, pp. x, 2, 12–4, 17) (property 42) through comprehensive general theories and models (ibid., pp. 8, 62, 169, 237, 400) (properties 46–47). Physics predicts and features predictive power (Gruenberger, 1964), it is fecund/fruitful (ibid.) with respect to predictions and to successful predictions, its community of theorists is concerned with generating correct predictions, and physics is useful (properties 42, 45, 50, 76, 82, and 184):

> Perhaps the most visible evidence of physics in everyday life is the eruption of new applications that have improved our quality of life, such as new medical devices, and advances in computers and high-tech communications.
>
> The exciting feature of physics is its capacity for predicting how nature will behave in one situation on the basis of experimental data obtained in another situation. Such predictions place physics at the heart of modern technology and, therefore, can have a tremendous impact on our lives. Rocketry and the development of space travel have their roots firmly planted in the physical laws of Galileo Galilei (1564–1642) and Isaac Newton (1642–1727). The transportation industry relies heavily on physics in the development of engines and the design of aerodynamic vehicles. Entire electronics and computer industries owe their existence to the invention of the transistor, which grew directly out of the laws of physics that describe the electrical behavior of solids. The telecommunications industry depends extensively on electromagnetic waves, whose existence was predicted by James Clerk Maxwell (1831–1879) in his theory of electricity and magnetism. The medical profession uses X-ray, ultrasonic, and magnetic resonance methods for obtaining images of the interior of the human body, and physics lies at the core of all these. Perhaps the most widespread impact in modern technology is that due to the laser. Fields ranging from space exploration to medicine benefit from this incredible device, which is a direct application of the principles of atomic physics. (Young & Stadler, 2018, pp. 1–2)

> The invention of the laser has been an extremely important factor in the rapid advances now taking place in optics, as well as many advances in the medical field [...]. The development of the infrared camera has provided a tool for the study of heat flow from buildings, which involves thermodynamics [...]. The rapid growth in consumer electronics, as seen in the availability of laptop computers, smartphones, and many other "essential" personal paraphernalia, has been made possible by developments in condensed-matter physics. These developments, as well as the development of photovoltaic solar cells [...], all involve applications of semiconductors. Particle physicists use particle accelerators to study the interactions of subatomic particles in high-energy collisions. The Large Hadron Collider [...] was used in the discovery of the Higgs Boson in 2012.
>
> Science and technology depend on each other for progress. Physics plays an important role in the education and work of engineers, whether they specialize in electrical, mechanical, nuclear, or other engineering fields. In fact, people with physics degrees often work as engineers when they are employed in industry. (Griffith & Brosing, 2022, p. 9)

Physics is communicable, as proved by all the works used in this chapter (property 49). Physicists do not pitch claims directly to the popular mass media first. It is scientifically peer-reviewed and does not evade scientific peer review (Gruenberger, 1964) (properties 127–128 and 208).

Physics has been and is progressive (Griffith & Brosing, 2022, pp. 14, 387, 404, 461, 470, 473) (properties 84, 176, and 213).

Physics features a high degree of general precision of formulation (Griffith & Brosing, 2022, p. 8) (property 89). It is mathematised (Margenau, 1935, pp. 52, 82; Mageshwaran et al., 2016, p. 27) (property 174). Its mathematisation involves the "[c]onstruction of […] formal theoretical framework[s]" (Grodzicki, 2021, pp. 51–2).

In physics, knowledge of facts is "obtained from observations; sometimes chance observations, but usually carefully-planned observations" (Lüders & Pohl, 2017, p. 3) (properties 51, 96, 102, 111–112, 147–148, and 154).

Physics deals with "reproducible events" (Grodzicki, 2021, p. 43) (properties 56–57). It is legal and offers explanations with respect to laws (ibid., pp. 212–23) (properties 181–182): "its laws encompass a wide scope of phenomena that range from the formation of galaxies to the interactions of particles in the nuclei of atoms" (Young & Stadler, 2018, p. 1). "Well-known examples are Newton's three laws of motion and the principle of the conservation of energy." (Frauenfelder & Huber, 1966, p. 1). Physics also explains using mechanisms (e.g., Griffith & Brosing, 2022, pp. 200, 206–10) (property 48).

Physics is empirically adequate within its intended domain of application (property 183) and enjoys a very high degree of theoretical confirmation (property 184). It is contiguous with many other current sciences (property 189), for example, astronomy, chemistry, and biology; and the domain of physics is included in those of these other sciences (for example, subatomic particles are present in chemical reactions, and are components of organisms):

> Physics is […] at the very root of subjects like chemistry, engineering, astronomy, and even biology. (Young & Stadler, 2018, p. 1)

> The principles and theories of physics can be used to explain the fundamental interactions involved in chemistry, biology, and other sciences at the atomic or molecular level. Modern chemistry, for example, uses the physical theory of *quantum mechanics* to explain how atoms combine to form molecules. Quantum mechanics was developed primarily by physicists in the early part of the twentieth century, but chemists and chemical knowledge also played important roles. Ideas about energy that arose initially in physics are now used extensively in chemistry, biology, and other sciences.
>
> […] In addition to physics, the physical sciences include chemistry, geology, astronomy, oceanography, and meteorology (the study of weather). Physics underlies all of them. (Griffith & Brosing, 2022, pp. 7–8)

This contiguity shows that physics is bulk externally consistent with current science (property 190).

Physics is relatively simple: "the models and ideas of physics can be described more simply and cleanly than those of other sciences." (Griffith & Brosing, 2022, p. 8) (property 188).

Physics looks for the composition, structure, and the kind and intensity of effects of things (property 197).

The international community of physicists features a high degree of disposition, willingness, and concern towards testing, confirming or disconfirming their theory. They are willing to test (e.g., Griffith & Brosing, 2022, p. 464) (property 206).

Physics studies "result in publicly verifiable knowledge" (property 64). It applies Ockham's razor (property 164), and is epistemically humble (property 198) and open-minded (property 199). It avoids the Fulton non sequitur/"Galileo fallacy"/"Galileo Gambit" (property 166). It recognises that it is epistemically limited and avoids the idea of having the correct explanation by default (properties 62 and 106) (Gruenberger, 1964).

14.3 The Properties That Physics (Mostly) Does Not Have

No property of science is indicated in the literature as clearly lacking in physics.

14.4 Chapter Conclusions

Physics has at least 29 of the specific properties of scientificity featured in the demarcatory list, while not clearly lacking any property. This yields a score of 29 for physics. This provisional tally stands strongly in favour of the scientificity of physics, and is the best explanation for the fact that physics is universally deemed scientific. This, once again, demarcates in a manner that is equivalent to that of the current international philosophical and scientific consensus regarding the extension of the term "science".

References

Frauenfelder, P., & Huber, P. (1966). *Introduction to physics*. Pergamon Press.

Griffith, W. T., & Brosing, J. W. (2022). *The physics of everyday phenomena: A conceptual introduction to physics* (10th ed.). McGraw-Hill.

Grodzicki, M. (2021). *Physical reality – construction or discovery? An introduction to the methodology and aims of physics*. Springer.

Gruenberger, F. J. (1964). A measure for crackpots. *Science, 145*, 1413–1415.

Lüders, K., & Pohl, R. O. (2017). *Pohl's introduction to physics* (Vol. 1, 2nd English ed.). Springer.

Mageshwaran, M., Sorli, A., & Fiscaletti, D. (2016). The foundations of the epistemology and the methodology of physics. *American Journal of Modern Physics. Special Issue: Insufficiency of Big Bang Cosmology, 5*(4–1), 27–33.

Margenau, H. (1935). Methodology of modern physics. *Philosophy of Science, 2*(2), 164–187.

Popper, K. (1959/2002). *The logic of scientific discovery* (Trans. *Logik der Forschung*. Repr.). Hutchinson.

Young, D., & Stadler, S. (2018). *Physics* (11th ed.). Wiley.

Chapter 15
The Properties of Chemistry

Abstract This chapter provides a brief definition and a description of/introduction to chemistry. It then meta-analytically shows that the demarcation criteria and the scoring method proposed in Chap. 2 demarcate chemistry as scientific, in agreement with the current philosophical and scientific consensus.

Keywords Chemistry · Scientific demarcation · Science · Scientificity

15.1 Introduction

Virtually everyone believes that current mainstream chemistry (from now on, referred to here simply by the term "chemistry") is a clearly scientific CF (e.g., cf. Baird et al., 2006; Scerri, 2008; Brown et al., 2012; Gabbay et al., 2012; Kovac & Weisberg, 2012). Chemistry exemplifies the practice of science in Popper's *The Logic of Scientific Discovery* (1959) and in his *Conjectures and Refutations* (1963) (and features in the works of prominent Popperians, e.g., Medawar). Furthermore, the chemical revolution is a prime example in Kuhn's *The Structure of Scientific Revolutions* (1996). Indeed, chemistry was central to the Scientific Revolution. *The Sceptical Chymist: or Chymico-Physical Doubts & Paradoxes* (1661) was one of the founding documents of that revolution. It was written by Robert Boyle, a chemist who helped separate chemistry from alchemy, from whose empirical knowledge chemistry had emerged (Hudson, 1992; Greenberg, 2000; Levere, 2001). Boyle made that separation while making chemistry more scientific and exhorting scholars to experiment (e.g., experiments with air pumps were central to the Scientific Revolution) when modern science was emerging. Chemistry was in its beginnings much more experimental than theoretical, and it was mathematised and systematised in laws later than physics (by, among others, Lavoisier, Cavendish, and Dalton).

This chapter shows that the demarcation criteria presented in Chap. 2 demarcate chemistry as scientific. This section defines "chemistry". Section 15.2 lists properties of scientificity that chemistry has. Section 15.3 lists properties of scientificity that chemistry lacks. As the conclusion of this chapter, Sect. 15.4 argues that the

D. Fernandez-Beanato, *A Working Scientific Demarcation*, Synthese Library 519, https://doi.org/10.1007/978-3-031-99005-2_15

demarcation criteria listed in Sect. 2.2 demarcate chemistry as scientific, and that the fact that virtually everyone believes that chemistry is scientific is explained by the fact that it has enough of the properties of science listed in Chap. 2.

The concern of this chapter is what properties it has and what properties it lacks. This chapter does not provide reasons for believing that chemistry is a science other than those that already exist in the literature. I take the properties of science listed in Sect. 2.2 and show that chemistry possesses many of them, according to the specialists in the context of discussing what chemistry is like. This provides: a) a rational reconstruction (and therefore an explanation) of *why* chemistry is scientific, b) further evidence of the extensional adequacy of this work, c) further evidence of this work's explanatory power, and d) further evidence of the fact that the proposed set of criteria demarcates science appropriately.

This assessment of chemistry uses scholarly literature. This chapter, like all other case studies in this work, incorporates analyses performed by several high-quality authors who are specialists in the field and who concluded that chemistry possesses or lacks properties that this work, in turn, uses for SD. I rely here on mainstream authoritative literature written by practising chemists and by philosophers of science. The texts used are key to this investigation because they engage with the properties of science that are featured in chemistry, *as* they are featured in chemistry. These attributions of properties are mutually commensurable (as part of mainstream traditions in the philosophy of science) and largely uncontroversial. They were arrived at by using different methods and are, in many cases, also independent of our problem, in the sense that the authors' concern when making them was not SD. Some of these works' authors are primarily chemists, while others are primarily philosophers. All of these authors are acknowledged authorities, and their works are standard, not controversial, with respect to the points taken up here. Also, many of these authors' works are rooted in detailed engagement with the current mainstream chemical science, which adds to their reliability. For all the above-mentioned reasons, we can rely on this set of authors for our purposes. Therefore, this work recognises that chemistry has the properties that those scientific or philosophical authorities adjudicated to it. In this way, this work is meta-analytical, in that its analysis uses and incorporates the conclusions of these authors' original analyses.

Nonetheless, explicit claims or arguments about the possession of properties of science by chemistry are difficult to find in the literature because the possession of these properties by chemistry is usually taken for granted, and that is why most of the properties listed in Chap. 2 are not covered in this chapter, although it seems clear that chemistry possesses most of them.

Our current understanding of what chemistry is depends on our current physical and chemical knowledge. A cursory exploration of the definition of "chemistry" usually yields the claim that chemistry is the study of the composition and behaviour of matter, or some similar claim; but this is obviously too general, as it does not demarcate chemistry from physics. A more specific definition is therefore required. Chemistry is the scientific CF specifically concerned with the transformations of the properties of matter that do not involve a change in atomic number or breaking down elementary forms of matter called elements, but only in the recombination of these elements. (This is our current theoretical understanding of what the kinds of paradigmatic reactions that have always been important in the history of chemistry have in common. This definition makes the extension of the term "chemistry" include work that was not done with knowledge of atomic numbers, and that therefore was done under a different understanding of what chemistry was, or even without a concept of "chemistry").

An element consists of a unique single type of atoms, characterised by the fact that they all have the same number of protons (the atomic number) in their nuclei. Chemistry is concerned with these elements, with compounds of these elements, and with mixtures of said elements or compounds.[1] "[A]ll matter in the Universe is composed of the atoms of these chemical elements" (Noller et al., 2020).[2] The main professional body in the CF of chemistry, the International Union of Pure and Applied Chemistry (IUPAC), currently (2025) recognises the existence of 118 chemical elements, of which 94 occur naturally, whilst the remaining 24 are synthetic, having been produced by humans.

"The fundamental principle of the science of chemistry is that the atoms of different elements can combine with one another to form chemical compounds." (Noller et al., 2020). A chemical compound is a chemical "substance composed of identical molecules[3] [(or molecular entities)] consisting of atoms of two or more chemical elements" (Noller et al., 2020) held together by chemical bonds. Chemical bonds are lasting attractions between atoms, ions, or molecules that enable the formation of chemical compounds. In mixtures, the constituent elements or compounds are not chemically combined by chemical bonds (IUPAC, 1997).

Chemistry is about the composition, structure, properties, and behaviour of these basic elements and compounds, and the changes they undergo during reactions. A chemical reaction is a chemical transformation of substances into one or more different substances, based on a rearrangement of electrons in the chemical bonds between their atoms (IUPAC, 1997). Chemistry is also centrally concerned with the properties of chemical bonds, and with phenomena such as how atoms and molecules interact via chemical bonds to form new chemical compounds.

Chemistry analyses the empirical world at a scale which is macro with respect to that of physics, and micro with respect to that of biology (the atom is the smallest entity that retains chemical properties of a chemical element, and therefore anything smaller is the purview of physics, not of chemistry). Chemistry connects the physical and biological aspects of the world, which is why it is often called "the central science".

Let us see how chemistry stands regarding the possession of our scientificity properties.

---

[1] Elements, compounds, and mixtures are composed of atoms, molecules, and ions.

[2] "Common usage might have us believe that 'chemicals' are just those substances [produced] in laboratories or something that is not a natural substance. Far from it, [...] everything is made of chemicals." (American Chemical Society, n.d.).

[3] A molecule is an electrically-neutral entity consisting of more than one atom held together by chemical bonds (IUPAC, 1997).

15.2 The Properties of Chemistry

Chemistry looks for correlations and strives to identify causes (property 23) (Hoyningen-Huene, 2013, pp. 98–102).

Chemistry offers (typically scientific) theoretical accounts (Hoyningen-Huene, 2013, p. 59), including explanations, and has explanatory power (Kuhn, 1996, p. 108; Early, 2003; Hoyningen-Huene, 2013, pp. 56–8) (property 42).

Chemistry generalises (Hoyningen-Huene, 2013, pp. 47–8). It "offers a multiplicity of models drawing on [...] experimental generalizations, [...] where appropriate." (properties 23 and 107) (Baird et al., 2006, pp. 4–5).

Chemistry predicts (i.e., it features predictive power) (property 45) and it is fecund/fruitful with respect to predictions and to successful predictions (it features predictive success) (property 82). For example, the publication of the periodic table of elements by Mendeleev in 1869 "enabled chemists [...] to predict the existence of yet undiscovered elements by deriving their properties from their position in the periodic table." (Hoyningen-Huene, 2013, p. 129).

Chemistry is reliably useful (property 50) (Gebelein, 1997):

> From its modern beginnings, not to mention its alchemical precursors, much of chemistry has been guided by a pragmatic orientation focused on the reliable manipulation and manufacture of substances. (Gabbay et al., 2012, p. 17)

> [O]n a larger scale, practical application of the natural sciences only emerged in the last quarter of the nineteenth century with the upcoming electrical industry and the chemical industry producing organic dyes. (Hoyningen-Huene, 2013, p. 117)

> One cannot help but be struck by the ubiquity of chemistry. Everything in sight has the chemist's art written into it. Paints, varnishes, and other finishes all are the products of chemistry, some long established and (in this sense) "low tech," some new "high tech" paints with nanoparticles for striking visual and other functional qualities. There are batteries, light bulbs—reminding us of Oliver Sacks' recent "memoir of a chemical childhood," Uncle Tungsten (Sacks' uncle made better light bulbs through chemistry)—and liquid crystal computer displays. There is even chemistry hidden inside silicon chips, which are made by using ultra-precise lithography. Many of us use cheap "gel writer" pens with their new improved ink. And, of course, everywhere one looks, one sees plastics. In the 1960s film "The Graduate," Dustin Hoffmann was given this advice about the future—"one word, plastics." In this new century, that word might be "nanomaterials." Either way, it is chemistry.
>
> Like the air we breathe, chemistry engulfs us. (Baird et al., 2006, p. 4)

> Chemistry greatly impacts our daily lives. Indeed, chemistry lies near the heart of many matters of public concern: improvement of health care, conservation of natural resources, protection of the environment, and provision of our daily needs for food, clothing, and shelter. Using chemistry, we have discovered pharmaceutical chemicals that enhance health and prolong lives. We have increased food production through the use of fertilizers and pesticides, and we have developed plastics and other materials used in almost every facet of our lives, from electronics to sporting equipment to building construction. (Brown et al., 2012, p. 111 (e-book version))

Chemistry is connected to (property 59), contiguous with (property 189), and bulk externally consistent (property 190) with the rest of science. "Chemistry plays an important part in all of the other natural sciences, basic and applied." (Russell, 1980, p. 2). "Chemistry offers a multiplicity of models drawing on [...] physical theory" (Baird et al., 2006, pp. 4–5), and is intimately continuous with biology in the chemical subfield of biochemistry. Chemistry imports contributions of other sciences whenever needed (Hoyningen-Huene, 2013, p. 140).

Chemistry features fecundity/fruitfulness, progress, and promise (properties 42 and 82–84):

> The discipline of chemistry is, in fact, enormous and enormously productive. [...] [T]he point [can be made] persuasively and concisely with data on the number of publications in various fields. With a grand total just shy of 900,000 papers indexed in chemical abstracts for the year 2000, chemistry is larger than all of the other natural sciences combined. [...] [C]ontemporary chemistry [(...) is], in fact, a very active and productive discipline [...] (Baird et al., 2006, p. 3)

Chemistry generates knowledge systematically (properties 93 and 178) (Hoyningen-Huene, 2013, p. 133).

Chemistry is mathematised (property 174) and analytical (properties 107 and 116):

> [Chemistry's reliable manipulation and manufacture of substances have] been realized through the conjoined methods of analysis, by which chemists determine the constituents of a given substance, and synthesis, by which predetermined substances, or more minimally, substances with desired properties, are produced. (Gabbay et al., 2012, p. 17)

Chemistry performs planned experimentation to achieve its cognitive goals (properties 111, 154, and 206). "[L]aboratory sciences like physics or chemistry often produce the phenomena they are interested in themselves. The obvious advantage is that they can get exactly the kind of data they want at the time they want them by producing the pertinent phenomena." (Hoyningen-Huene, 2013, pp. 135–6).

Chemistry successfully uses diverse procedures for the elimination of errors (properties 148, 209, 214, and 217) (Hoyningen-Huene, 2013, pp. 94–8).

Chemistry provides legal explanations (properties 181–182): "[M]odels used for explanation are typically models about the system's composition, together with [...] laws of the interaction of these components" (Hoyningen-Huene, 2013, p. 66).

The chemical community is specialised (property 201) (Hoyningen-Huene, 2013, p. 194).

15.3 The Properties That Chemistry (Mostly) Does Not Have

None of the properties of science seems to be clearly lacking in chemistry.

Regarding all the properties in Sect. 2.2 not mentioned in this chapter, it is clear that chemistry mostly has most of them. Explicit claims or arguments in this regard are difficult to find in the literature precisely because the possession of these properties by chemistry is usually taken for granted.

15.4 Chapter Conclusions

Chemistry has at least 17 of the specific properties of scientificity featured in the demarcatory list, while not clearly lacking any property. This yields a score of 17 for chemistry. This (possibly provisional) tally stands strongly in favour of the scientificity of chemistry, and is the best explanation for the fact that chemistry is commonly deemed scientific. This, once again, demarcates in a manner that is equivalent to that of the current international philosophical and scientific consensus regarding the extension of the term "science".

References

American Chemical Society. (n.d.). *Chemistry is everywhere*. https://www.acs.org/content/acs/en/education/whatischemistry/everywhere.html

Baird, D., Scerri, E., & McIntyre, L. (Eds.). (2006). *Philosophy of chemistry: Synthesis of a new discipline*. Springer.

Brown, T. L., LeMay, H. E., Jr., Bursten, B. E., Murphy, C. J., & Woodward, P. M. (2012). *Chemistry: The central science*. Prentice Hall.

Earley, J. E., Sr. (Ed.). (2003). *Chemical explanation: Characteristics, development, autonomy*. New York Academy of Sciences.

Gabbay, et al. (2012). *Handbook of the philosophy of science, Vol. 6: Philosophy of chemistry*. Elsevier.

Gebelein, C. G. (1997). *Chemistry and our world*. William C. Brown.

Greenberg, A. (2000). *A chemical history tour: Picturing chemistry from alchemy to modern molecular science*. Wiley.

Hoyningen-Huene, P. (2013). *Systematicity: The nature of science*. Oxford University Press.

Hudson, J. (1992). From alchemy to chemistry. In J. Hudson (Ed.), *The history of chemistry* (pp. 35–46). Springer.

IUPAC. (1997). *Compendium of chemical terminology* (2nd ed.) (the "Gold Book"). Compiled by A. D. McNaught and A. Wilkinson. Blackwell Scientific Publications. Online version (2019–) created by S. J. Chalk. https://goldbook.iupac.org/

Kovac, J., & Weisberg, M. (Eds.). (2012). *Roald Hoffmann on the philosophy, art, and science of chemistry*. Oxford University Press.

Kuhn, T. S. (1996). *The structure of scientific revolutions* (3rd ed.). University of Chicago Press.

Levere, T. H. (2001). *Transforming matter: A history of chemistry from alchemy to the buckyball*. JHU Press.

Noller, C. R., Norman, R. O. C., Zumdahl, S. S., & Usselman, M. C. (2020). Chemical compound. *Encyclopedia Britannica*. https://www.britannica.com/science/chemical-compound

Russell, J. B. (1980). *General chemistry*. McGraw-Hill International.

Scerri, E. (2008). *Collected papers on philosophy of chemistry*. Imperial College Press.

Chapter 16
The Properties of Astronomy

Abstract This chapter provides a brief definition and a description of/introduction to astronomy. It then meta-analytically shows that the demarcation criteria and the scoring method proposed in Chap. 2 demarcate astronomy as scientific, in agreement with the current philosophical and scientific consensus.

Keywords Astronomy · Scientific demarcation · Science · Scientificity

16.1 Introduction

Virtually everyone believes that current academic astronomy is scientific. Astronomy is one of the oldest sciences in human culture, and an observational and mature natural science (Kuhn, 1996, p. 69; Unsöld & Baschek, 2002, p. 1; Dietrich, 2011; Hoyningen-Huene, 2013, pp. 47, 91). This section provides a working definition of the term "current academic astronomy". Section 16.2 lists properties of science that current academic astronomy has. Section 16.3 lists properties of science that it lacks. As the conclusion of this chapter, Sect. 16.4 argues that the demarcation criteria listed in Sect. 2.2 demarcate current academic astronomy as scientific, and that the fact that virtually everyone believes that current academic astronomy is scientific is explained by the fact that it has enough of the properties of science listed in Chap. 2.

Current academic astronomy is a CF that studies and explains celestial objects and phenomena. It is "the study of objects and matter outside the Earth's atmosphere and of their physical and chemical properties" (Merriam-Webster). This includes planets and their satellites, asteroids, comets, and meteoroids, interstellar matter and radiation occurring in space; systems of stars, star clusters, dark matter, gas, and dust called "galaxies"; and galaxy clusters; their formation, structures, positions, movements, evolution, and linked phenomena.

As with all scientific CFs, an astronomical field is scientific as long as it possesses enough of the properties of scientificity and is the one that contextually (in a given period and society) has the most of them (in this way, previous developments

D. Fernandez-Beanato, *A Working Scientific Demarcation*, Synthese Library 519, https://doi.org/10.1007/978-3-031-99005-2_16

stop being scientific as soon as a (demonstrably) sufficiently epistemically-progressive integral separated approach develops).

From now on, this chapter uses the term "astronomy" to refer to current academic astronomy.

The concern of this chapter is what properties of science astronomy has and what properties of science it lacks. This chapter does not provide reasons for believing that astronomy is a science other than those that already exist in the literature. I take the properties of science listed in Sect. 2.2, show that astronomy possesses many of them, and that this is asserted by the specialists on astronomy in the context of discussing what astronomy is like. This provides: a) a rational reconstruction (and therefore an explanation) of *why* astronomy is scientific, b) further evidence of the extensional adequacy of this work, c) further evidence of this work's explanatory power, and d) further evidence of the fact that the proposed set of criteria demarcates science appropriately.

This work assesses astronomy by using scholarly literature. This chapter, like all other case studies in this work, incorporates analyses performed by high-quality authors who concluded that astronomy possesses or lacks properties that this work, in turn, uses for SD. I rely here on mainstream authoritative literature on astronomy and on the philosophy of astronomy written by philosophers of science. The texts used are key to this investigation because they engage with the properties of science that are featured in astronomy, *as* they are featured in astronomy. These attributions of properties are mutually commensurable (as part of mainstream traditions in the philosophy of science) and largely uncontroversial. They were arrived at by using different methods and are, in many cases, also independent of our problem, in the sense that the authors' concern when making them was not SD. All of these works are sufficiently up-to-date, and are standard, not controversial, with respect to the points taken up here. For all these reasons, we can rely on this set of authors for our purposes. Therefore, this work recognises that astronomy has the properties that those scientific or philosophical authorities adjudicated to it. In this way, this work is meta-analytical, in that its analysis uses and incorporates the conclusions of these authors' original analyses.

Nonetheless, explicit claims or arguments about astronomy's possession of properties of science are difficult to find in the literature because the possession of these properties by astronomy is usually taken for granted, and that is why most of the properties listed in Chap. 2 will not be covered in this chapter, although it seems clear that astronomy possesses most of them.

Let us see how astronomy stands regarding the possession of our scientificity properties.

16.2 The Properties of Astronomy

Astronomy generalises (properties 23, 32, and 107). "Typical generalizing empirical sciences [...] include some parts of observational disciplines like astronomy, e.g., where generalized descriptions of stellar evolution are sought" (Hoyningen-Huene, 2013, p. 47).

Astronomy predicts (i.e., it features predictive power), it is fecund/fruitful with respect to predictions and to successful predictions (Brush, 1995), its community of theorists is concerned with generating correct predictions, and astronomy is useful (properties 42, 45, 50, 76, 82, and 184):

> The most obvious and stringent regularities, accessible and of interest to all cultures of all ages, have been and still are astronomical regularities concerning the (apparent) motion of celestial bodies, most notably the sun and the moon. Although knowledge about the regular change of day and night, for instance, can certainly not count as scientific knowledge, [astronomy is about] knowledge about the more sophisticated regularities of phenomena like solar or lunar eclipses [...]. [...] A classic case [of successful astronomical prediction] is the discovery of the planet Neptune in 1846. Earlier, discrepancies between calculations and data for the orbit of the planet Uranus had been detected. These calculations were based on Newton's theory of gravitation. A possible explanation for these discrepancies was the existence of a yet undiscovered planet beyond Uranus that gravitationally influenced its orbit. The British mathematician John Couch Adams and the French mathematician and astronomer Urbain Le Verrier independently worked out this hypothesis. On the basis of the observed discrepancies and Newton's gravitational theory, they were able to predict the position of this yet unknown planet. On September 23, 1846, the new planet was indeed detected by the German astronomer Galle at the Berlin Observatory, almost exactly at the position predicted by the calculations. (Hoyningen-Huene, 2013, pp. 80, 84)

> A part of normal theoretical work, though only a small part, consists simply in the use of existing theory to predict factual information of intrinsic value. The manufacture of astronomical ephemerides [...] are examples of problems of this sort (Kuhn, 1996, p. 30)

Astronomy is falsifiable (property 54), is concerned with falsifiability when constructing its theories; and is precise (property 89) and clear (property 92):

> John Mather and John Boslough, historians of astronomy, link Popper's notion of falsifiability to many significant advances in astronomical theory construction. [...] These authors remarked that this type of thinking was the determining factor in the acceptance of cosmic background radiation and the big bang theory, two of the greatest astronomical discoveries [(...) of the 20th century]. These were winners for the simple reason that the current models could not reasonably account for the data being observed. Thus, Popper's concepts of falsification and precision determined which astronomical theories were fit to survive. There is further evidence that Popper has begun to guide how modern astronomers construct their theories. Bruce Balick and Adam Frank, astronomers at the Universities of Rochester and Washington, respectively, noted that modern-day astronomers construct theories with clear predictions meant to be easily disproved. These authors reported that the Hubble Space Telescope's observations of penetrating spatial resolution falsified the theory of planetary nebulae, and that stunning images from Hubble and the Chandra X-ray satellite [now called the Chandra X-ray Observatory] have put in doubt longstanding theories of stellar evolution, mass transfer in binaries, and effects of putative magnetic fields (Balick & Frank, 2002, pp. 439-440). They remarked that the reasons these theories were so easily falsified

related to the notion of clarity and testability in their construction. (Sovacool, 2005, pp. 55–6)

Astronomy is systematic and systematically gathers information (properties 95, 178, 193, and 195):

The beginning of the science of astronomy is marked by systematic observation of celestial phenomena that were known beforehand, and their systematic descriptions. For instance, early astronomy produced catalogs of stars, descriptions of striking astronomical processes, the discovery of periodicities in the latter, and the like. In this way, common sense knowledge was made more precise, in the form of catalogs or even with the help of mathematics. This increase in systematicity of astronomical knowledge [...] (Hoyningen-Huene, 2013, p. 191)

[M]ost empirical sciences are constantly on the move to systematically improve existing data and to gain new ones. [...] [S]cience did not settle for unaided observation, and the history of astronomy is a particularly good case in point. Already before the invention of the telescope, various instruments were in use, mainly aiding the measurement of the position of celestial objects. With the invention of the optical telescope and its use in science from 1609 on by Galileo Galilei, however, a new era of collecting data about astronomical objects had begun. Immediately from its invention on, the optical telescope was continuously and systematically improved, and already Galileo had started this development. Many new designs of the optical telescopes have been invented and continue to be invented. Ever since 1609, the optical telescope has been permanently used in astronomy. In the beginning, its light-gathering power was something like ten times that of the unaided human eye, corresponding to an objective lens diameter of roughly two centimeters. Four hundred years later, the most powerful optical telescopes feature mirrors equivalent to an aperture of almost twelve meters, corresponding to a light-gathering power three million times more than the human eye. There are even plans to build gigantic optical telescopes exceeding the power of today's largest telescopes by a factor of up to one hundred. In addition to the optical telescope operating with visible light, other sorts of telescopes were invented, systematically using all other media like X-rays, infrared radiation, or radio waves to gather information about astronomical objects. In addition, space-based telescopes of various sorts systematically gather data of extreme importance to cosmology. For instance, the Wilkinson Microwave Anisotropy Probe (WMAP) is a NASA explorer mission dedicated to measuring the cosmic microwave background radiation, especially its anisotropy (the radiation's uneven distribution over the sky). In 2001, the measuring instrument was put on a spacecraft and put in an orbit at a distance of 1.5 million kilometers from the Earth. Until 2010, it measured data of unprecedented accuracy and highest relevance to cosmology. For instance, from its data, the age of the universe was determined to be 13.73 billion years within 1 percent.

Not only were astronomical instruments systematically improved, but also their use is systematic and has become even more systematic. Just one topical example is the Sloan Digital Sky Survey. It uses an Earth-based 2.5-meter telescope and is one of the most ambitious and influential surveys in the history of astronomy. It started in 2000, and within its first eight years of operation, it systematically covered one quarter of the entire sky. It created three-dimensional multicolor images that contain more than 930,000 galaxies and 120,000 quasars. Furthermore, our own galaxy was investigated, and the spectra of 240,000 of its stars were recorded. Finally, systematic surveys of variable objects on the sky discovered some five hundred super-novae of a particular type that are important in determining details of the cosmic expansion over the last four billion years. The data produced by the Sloan Digital Sky Survey have been used in a wide range of astronomical investigations around the world, and, as of September 2008, they have contributed to more than two thousand articles published in scientific journals. (Hoyningen-Huene, 2013, pp. 134–5)

Astronomy is thoroughly mathematised (property 174) and specialised (property 201): "[I]n [...] astronomy, research reports had ceased already in antiquity to be intelligible to a generally educated audience." (Kuhn, 1996, p. 20). Astronomy is empirically adequate within its intended domain of application (property 183) and enjoys a very high degree of theoretical confirmation (property 184). It is contiguous with other current sciences (property 189), especially physics; and each member of the domain of one of these fields is a component of a system in the domain of the other (for example, both particles and chemical bonds are components of celestial objects):

> Astronomy [...] deals with the quantitative investigation of the cosmos and the physical laws which govern it [...]. In the 19th and particularly in the 20th centuries, physics assumed the decisive role in the elucidation of astronomical phenomena; astrophysics has steadily increased in importance over "classical astronomy". There is an extremely fruitful interaction between astrophysics/astronomy and physics: on the one hand, astronomy can be considered to be the physics of the cosmos, and there is hardly a discipline in physics which does not find application in modem astronomy; on the other hand, the cosmos with its often extreme states of matter offers the opportunity to study physical processes under conditions which are unattainable in the laboratory. Along with physics, and of course mathematics, applications of chemistry and the Earth and biological sciences are also of importance in astronomy. (Unsöld & Baschek, 2002, p. 1)

This contiguity shows that astronomy is bulk externally consistent with current science (property 190).

16.3 The Properties That Astronomy (Mostly) Does Not Have

It is usually agreed that one shortcoming of astronomy as a science is that it is experimentally limited, in that it cannot experimentally manipulate much of what it studies, e.g., celestial bodies as such (Unsöld & Baschek, 2002, p. 1; Morison, 2008, p. 1). It is possible to perform experiments on entities from the astronomical realm, e.g., starlight and moonrocks, but those experiments pertain to the domains of other sciences, e.g., physics, optics, chemistry, and geology, and are not astronomical experiments. Nonetheless, experimentation is not necessary for the possession of any of our properties.

Regarding all the properties in Sect. 2.2 not mentioned in this chapter, it is clear that astronomy possesses most of them.

16.4 Chapter Conclusions

Astronomy has at least 13 of the specific properties of scientificity featured in the demarcatory list, while not clearly lacking any property. This yields a score of 13 for astronomy. This (possibly provisional) tally stands strongly in favour of the scientificity of astronomy, and is the best explanation for the fact that astronomy is commonly deemed scientific. This, once again, demarcates in a manner that is equivalent to that of the current international philosophical and scientific consensus regarding the extension of the term "science".

References

Balick, B., & Frank, A. (2002). Shapes and shaping of planetary nebulae. *Annual Review of Astronomy and Astrophysics, 40*, 439–486.

Brush, S. G. (1995). Prediction and theory evaluation in physics and astronomy. In A. J. Kox & D. M. Siegel (Eds.), *No truth except in the details. Boston studies in the philosophy of science* (Vol. 167). Springer.

Dietrich, T. K. (2011). *The culture of astronomy: Origin of numbers, geometry, law, science, and religion*. Bascom Hill Publishing Group.

Hoyningen-Huene, P. (2013). *Systematicity: The nature of science*. Oxford University Press.

Kuhn, T. S. (1996). *The structure of scientific revolutions* (3rd ed.). University of Chicago Press.

Morison, I. (2008). *Introduction to astronomy and cosmology*. Wiley.

Sovacool, B. (2005, February). Falsification and demarcation in astronomy and cosmology. *Bulletin of Science, Technology & Society of Virginia Polytechnic Institute and State University*. https://journals.sagepub.com/doi/pdf/10.1177/0270467604270151?casa_token=eq2khRdH-M0kAAAAA:d08nu215FLS7oTo3zoqtIdQ2pujXFKCbdTWCxj1GbAMWPaemk5EoS8hDTxNlpVnZ2aOwww68RdTi

Unsöld, A., & Baschek, B. (2002). *The new cosmos: An introduction to astronomy and astrophysics* (5th ed., corrected 2nd printing). Springer.

Chapter 17
The Properties of Biology

Abstract This chapter provides a brief definition and a description of/introduction to biology. It then meta-analytically shows that the demarcation criteria and the scoring method proposed in Chap. 2 demarcate biology as scientific, in agreement with the current philosophical and scientific consensus.

Keywords Biology · Scientific demarcation · Science · Scientificity

17.1 Introduction

Virtually everyone believes that current academic biology (CAB) is scientific. This section defines the term "CAB". Section 17.2 lists properties of science that CAB has. Section 17.3 lists properties of science that CAB lacks. As the conclusion of this chapter, Sect. 17.4 argues that the demarcation criteria listed in Sect. 2.2 demarcate CAB as scientific, and that the fact that virtually everyone believes that CAB is scientific is explained by the fact that it has enough of the properties of science listed in Chap. 2.

As a CF and a discipline, CAB is clearly scientific (Curtis et al., 2014, pp. 4–5). The concern of this chapter is what properties it has and what properties it lacks. This chapter does not provide reasons for believing that CAB is a science other than those that already exist in the literature. I will take the properties of science listed in Sect. 2.2, show that CAB possesses many of them, and that this is asserted by the specialists on CAB in the context of discussing what CAB is like. This provides: a) a rational reconstruction (and therefore an explanation) of *why* CAB is scientific, b) further evidence of the extensional adequacy of this work, c) further evidence of this work's explanatory power, and d) further evidence of the fact that the proposed set of criteria demarcates science appropriately.

It would be impossible to conduct an in-depth study of the whole of CAB in this chapter, but this work does not depend on contentious attributions of properties to fields: assessing CAB by using the scholarly literature will be sufficient for our purposes. This chapter, like all other case studies in this work, incorporates analyses

D. Fernandez-Beanato, *A Working Scientific Demarcation*, Synthese Library 519, https://doi.org/10.1007/978-3-031-99005-2_17

performed by several high-quality authors who are specialists in the field and who concluded that CAB possesses or lacks properties that this work, in turn, uses for SD. I rely here on mainstream authoritative literature on CAB and on the philosophy of biology written by practising biologists and by philosophers of science. The texts used are key to this investigation because they engage with the properties of science that are featured in CAB, *as* they are featured in CAB. These attributions of properties are mutually commensurable (as part of mainstream traditions in the philosophy of science) and largely uncontroversial. They were arrived at by using different methods and are, in many cases, also independent of our problem, in the sense that the authors' concern when making them was not SD. Some of these works are more "classic" and some are more recent, but all of them are sufficiently up-to-date. Some of their authors are primarily biologists, while others are primarily philosophers, some being accomplished in both disciplines. All of these authors are acknowledged authorities about CAB, and their works are standard, not controversial, with respect to the points taken up here. Also, all of these authors' works are rooted in detailed engagement with the current academic biological science, which adds to their reliability. For example, Weber was a practising biologist before becoming a philosopher of science. Consequently, this chapter draws especially upon his *Philosophy of Experimental Biology*. I also make heavy use of Darden's *Theory Change in Science*, because it particularly considers CAB's methods and techniques, which are one of the dimensions of science that are key for SD. For all the above-mentioned reasons, we can rely on this set of authors for our purposes. Therefore, this work recognises that CAB has the properties that those scientific or philosophical authorities adjudicated to it. In this way, this work is meta-analytical, in that its analysis uses and incorporates the conclusions of these authors' original analyses.

Nonetheless, explicit claims or arguments about CAB's possession of properties of science are difficult to find in the literature because the possession of these properties by CAB is usually taken for granted, and that is why most of the properties listed in Chap. 2 will not be covered in this chapter, although it seems clear that CAB possesses most of them.

Biology is the study of, and theory on, the phenomenon of life,[1] life processes, or living things.[2] Therefore, all other life sciences are sub-disciplines of biology.

CAB is a highly sophisticated and successful CF. It is the outcome of the epistemic progress of biology through time, and currently features very stable core theories. It includes a multitude of subject matter areas, pertaining to different topics or levels of aggregation, and encompasses corresponding subfields: it includes matters of origin, biochemistry, history, reproduction, inheritance, genetics, interaction, evolution, structure and form (morphology: eidonomy, anatomy, histology),

[1] The definition of "life", on which the definition of "biology" depends, is a more contentious matter (Tsokolov, 2009), but for our purposes here we do not need to get into that debate.

[2] Treviranus (1802, p. 4) defined (scientific) biology as "the science that concerns itself with" "the different forms and manifestations of life, the conditions and laws under which these phenomena occur, and the causes through which they have been affected.".

functions (physiology, neuroscience), growth, behaviour (ethology), ways of life, development (including embryology and biological gerontology), distribution, taxonomy, ecology, immunology, pathology, epidemiology, etc.

As a CF and a discipline, CAB is a mature science. (There have been, and there can be, non-scientific biologies, including prescientific, folk, and pseudoscientific views of life, which may include prescientific, folk, or pseudoscientific medicines,[3] but all of these, as they are not part of CAB, will not concern us in this chapter.) As with all scientific CFs, a biological field is scientific only as long as it possesses enough of the properties of science and is the one that contextually (in a given period and society) has the most of them (in this way, previous developments stop being scientific as soon as a (demonstrably) sufficiently epistemically-progressive integral separated approach develops). In this sense, CAB is the most scientifically advanced biology we humans have.

Although biology has been consolidated as a science (i.e., a scientific CF) for approximately only two centuries, human biological thought and knowledge are thousands of years old. Biological knowledge going beyond pure common sense has been cemented as part of human knowledge for thousands of years (e.g., the knowledge that bees pollinate vegetable species, which dates from Classical Antiquity). Until the Nineteenth century, human non-medicinal biological knowledge was largely composed of descriptions of observable features, and of taxonomies. This changed with the further realisations of common descent, natural selection, genetic heredity, and biological evolution,[4] which together constituted a profound conceptual revolution in biology. The modern proposal of evolution (firstly called "transformation"), starting with Lamarck in 1800, and the rediscovery of Mendelian genetics, in 1900, were unified in the so-called Modern Evolutionary Synthesis in the first half of the Twentieth century.

From now on, this chapter will use the term "biology" to refer exclusively to CAB.

Let us see how biology stands regarding the possession of our scientificity properties.

---

[3] There has been some debate on whether medicine is only a part of biology, an application of biology, or a different separate discipline. This chapter will not discuss medicine, except insofar as it is a part of, or overlaps with, biology. Medical biology is all biology that has applications in medicine, health care, or laboratory diagnostics. While there are many unproven and pseudoscientific ideas and practices that are presented as medicinal (including some so-called "alternative" or "complementary" medicines), all medicine that is scientific is based on CAB. Pseudoscience is prevalent in medicine due to several reasons; in particular, because of the existence of a market for alleged cures and palliatives. While there are no rivals to CAB or to scientific medicine in equal epistemic standing, there are non-sciences that challenge parts of CAB insofar as they promote rival (inconsistent) non-scientific ideas about centrally biological topics such as physiology, etc. Some of them might have been protoscientific medicines at some point in time (and, at times, more epistemically advanced than Western medicines), in the same way in which feng shui might have been protoscientific in the past, but they are non-scientific now, which is why currently advocating them is a pseudoscientific attitude (see Chap. 7). See Chap. 6 for the pseudomedicine of homeopathy.

[4] In addition to pseudomedicines, another pseudoscientific challenge to current scientific biology comes from creation "science", covered in Chap. 5.

17.2 The Properties of Biology

Because biology features properties of all the different types mentioned in our demarcatory list in Sect. 2.2, this section differs from the equivalent sections in other chapters by listing properties of biology under their types.

* Non-specific properties related to practices, methods, and techniques (specifically, properties 1, 7 (which requires property 54), 8–9, 13, 17, and 23): Biology does theoretical findings by combining data with educated guesses (hypotheses) (Darden, 1991, p. 224; Sousa, 2016, p. 110). It discriminates hypotheses (Reiser et al., 2013, p. 264) by evaluating them against this data. It features confirmations and disconfirmations (see Darden, 1991, pp. 58, 79, 130–2, 137–8, 150, 159, 162, 164–7, 203, 212, 223, 239, 263, 302; Weber, 2005, pp. 78, 174–5, e.g., pp. 64, 72, 74, 112–3, 157, 203, 217, 219, 275, 277, 285, 291) and rational debate (see Darden, 1991; Weber, 2005). In biology, lack of ad hocness is an "[e]specially important criteri[on] [...]. It is important that [...] new theoretical component[s] [...] not be merely [...] ad hoc addition[s] that serve[...] only to account for [...] anomal[ies]." (Darden, 1991, pp. 211, 217). Biology considers its disconfirmations, misses, and incorrect predictions (Darden, 1991, pp. 159, 164–5, 174, 203). For example,

> [l]ike most biological theories, the theory of classical genetics was not created by a single stroke of genius. Instead, it originated from a very simple precursor by a series of modifications triggered by experimental anomalies. [...] [T]hese anomalies necessitated revisions in genetic theory. [...] [A] number of [...] anomaly-driven revisions [...] occurred between 1900 and 1926. (Weber, 2005, pp. 8–9, see pp. 63–72 for more details, more examples in Chapter 3)

Biology leaves room for speculations to be superseded (e.g., Darden, 1991, p. 145). For example,

> [a]fter the rediscovery of Mendel's work in 1900, there was a notable trend to move away from the highly speculative and metaphysically loaded notions of "gemmules," "pangenes," or "idioblasts" (a term introduced by August Weismann) that were so popular in nineteenth-century biological thinking. (Weber, 2005, p. 195)

Biology looks for, and explains, correlations and patterns (Sousa, 2016, p. 111; Voit, 2019, p. 5; e.g., Darden, 1991, pp. 100, 137; e.g., Weber, 2005, pp. 59, 141, 285, 293), and uses correlation reasoning/thinking. It applies statistical methods to develop "descriptive studies, such as the characterization of the genome, proteome, and metabolome of model species or of an individual, as well as the characterization of a pathological or environmental change condition." (Sousa, 2016, p. 111).

* Non-specific formal properties of propositions, hypotheses, theories, and arguments (specifically, property 31): Biology reasons both deductively and non-deductively (Weber, 2005, pp. 180–7, 305, 309; Sousa, 2016, p. 109; Voit, 2019, pp. 1, 5; e.g., Weber, 2005, pp. 62–3). It formulates hypotheses (Darden, 1991, p. 224; Reiser et al., 2013, p. 264).

* Non-specific properties related to the contents of problems, propositions, hypotheses, and theories (semantical aspects, metaphysical aspects, aboutness, etc.) (specifically, properties 39, 42, 45–46, 48, 51, 53–54, 56–57, 59–60, and 63):

Biology's general philosophical views contain a view of science as organised activity which aims at least in part at factually truthlike descriptions (Weber, 2005, p. 78). It offers corroborated descriptions, corroborated explanations, and models (Darden, 1991, p. 245; Weber, 2005, p. 19 ("biology today accomplishes exactly this; namely, it explains certain biological phenomena"); Voit, 2019, pp. 6–9). It has the capability to be used to explain (and has power to explain) observable phenomena (Scriven, 1959, p. 477; Darden, 1991, pp. 100, 240; Weber, 2005, e.g., pp. 72, 112–3, 217–8, 285; Sousa, 2016, pp. 110–1). Biology predicts (Darden, 1991, pp. 29–30, 58, 137–8, 141, 143, 159–62, 164–6, 178, 224, 234–5, 238–40, 245, 283, 296, 299, 300–1; Weber, 2005, pp. 67, 96–7, 99–103, 111–3, 121, 126, 183, 218, 277, 284, 303, 314). It offers mechanistic explanations (Darden, 1991, Part 1; Bechtel & Richardson, 1993; Weber, 2005, pp. 7, 19, 21–4, 26–7, 30, 37, 42–3, 47–8, 56, 63, 76, 78, 84–5, 93, 95, 108, 114, 125, 135–7, 141–2, 145, 153–4, 156, 159, 180–6, 196, 202, 234–6, 242, 248–9, 255–8, 263–5, 270–1, 277, 301–2, 305, 309, 313–4, 317–8). Its propositions are non-tautologous, contingent, and testable (Darden, 1991, pp. 15, 224; Weber, 2005, Chapter 4), as there is experimentation in biology (Darden, 1991, p. 296; Sousa, 2016, pp. 110–1; Weber, 2005, chapters 3–4, e.g., p. 154: "During the second decade of the twentieth century, Thomas Hunt Morgan and his students laid the foundations of modern genetics by experimenting with just a single species, the fruit fly *Drosophila melanogaster*"). Biology features theoretical boundary conditions (Brown, 1994, pp. 142–3; Weber, 2005, pp. 25–6). As an empirical science, biology features theoretical in-principle falsifiability, which is why it can feature confirmations and disconfirmations. Famously, Popper denied the falsifiability of evolutionary biology, which is one of the foundational theories of current biology, but he later recanted that position, indicating: "I have changed my mind about the testability and logical status of the theory of natural selection" (Popper, 1978, quotation: p. 345; Sousa, 2016, p. 112; see also Scriven, 1959; and Mayr, 2004, pp. 70–4). Observations of phenomena in biology are repeatable/replicable (Darden, 1991, pp. 42, 50; Weber, 2005, pp. 116–8) and have been repeated/replicated (e.g., Weber, 2005, p. 194). Biology is unified (Mayr & Provine, 1980; Smocovitis, 1992). "[T]he struggle to unify the biological sciences was over by 1955". "By 1955 biology had become not only a unified science [...], but a mature science secure of its foundations" (Smocovitis, 1992, pp. 55, 57). What is more: in the 1950s and 1960s, "[a]s research in molecular biology and biochemistry intensified, the links between physicists and chemists and biologists solidified further." (Smocovitis, 1992, p. 58). Now, biology is "prevalently" connected with the rest of current science, especially with physics and chemistry (Darden, 1991, pp. 65, 132):

> Much of the progress of modern biology results from the development of interfield theories and the progressive unification (i.e., bridging) of the biological and physical sciences. With the chromosome theory, genes were associated with intracellular organelles; subsequently, with the development of the Watson–Crick model, the genes were identified with segments of DNA molecules with a specific type of structure. The operon theory provided further links between genetics and bio-chemistry, and the theory of allosteric regulation served as a bridge between biochemistry and physical chemistry. In sum, the chromosome theory was the first of a series of interfield theories, the operon theory and theory of allosteric regula-

> tion among them, that advanced our understanding of the relationship between the biological and physical sciences and resulted in progress in modern biology. (Darden, 1991, pp. 145–6)

In biology, generalisations that "unify a large number of otherwise disparate phenomena" "actually exist, [...] in areas such as evolutionary theory." (Weber, 2005, p. 45). The information provided by biology is fallible and tentative. It should be deemed only probable, and therefore provisional. It is also not "about final—or ultimate—answers" (Sousa, 2016, p. 115; quote: Lack & Rousseau, 2016, pp. 22–3, 31, 47).

* Non-specific properties related to epistemic attitudes and values (axiological properties), individual or social (specifically, properties 68, 70, 73–74, 76, and 79 (which requires properties 51–52)): Biology formulates hypotheses (Darden, 1991, p. 224; Reiser et al., 2013, p. 264) and other types of accounts (descriptions, explanations, models); therefore, it is a creative activity. It strives for knowledge of causal mechanisms (Darden, 1991, Part 1, Chapter 7, and Part 3). It recognises the limitations of speculative theorising (e.g., Darden, 1991, p. 145). Its community of theorists is concerned with reducing the number of unexplained phenomena (e.g., Darden, 1991). Regarding the evaluation of theories, models, or explanations, for the biology research community it is an "evaluative demand" that "plausible rivals be investigated", and the "consideration of rivals" is an "evaluative strategy" (assessment criterion) (including for the solution of anomalies) (Darden, 1991, pp. 215, 272, 289–91, 294, 299–300, 305–6; e.g., p. 304). This helps create plausible hypotheses (Darden, 1991, p. 294). Predictive adequacy is also an evaluative strategy for theories in biology (Darden, 1991, pp. 217, 300). Biology shows regard of refuting information (Darden, 1991, pp. 159, 164–5, 174, 203).

* Non-specific epistemic properties of practices, propositions, hypotheses, and theories, and of the evolution of the discipline (specifically, properties 81–82 and 84 (which requires property 42)): In biology, success is achieved through progressive unification. For example, "the independent assortment of genes (Mendel's second law) was explained, according to Kitcher, by instantiating a pairing and separation schema, thereby showing that chromosomal pairing and separation was a unifying natural kind." (Kitcher, 1984; quote: Darden, 1991, p. 107). There is also a substantial amount of progressive unification with other scientific CFs (see above in this section). This unification is fecund/fruitful. For example, "[p]rogress occurred in theory construction in Mendelian genetics when the appeal to vague analogies [used by William Bateson] was replaced by fruitful interfield connections to cytology" "in T. H. Morgan's development of the theory of the gene" (Darden, 1991, pp. xiii, 5, 138):

> [T]he chromosome theory of Mendelian heredity is an interfield theory bridging the fields of genetics and cytology. It was generated to unify the knowledge of heredity found in both fields and thereby to explain the similar properties of chromosomes and genes. It functioned to focus attention on previously neglected items of the domains and to predict new items for the domains of each field. It further served to generate a new line of research coordinating the fields of cytology and genetics. Success in finding the predictions of the theory and in developing the common line of research resulted in the confirmation of the theory and the fruitful bridging of two fields of science. (Darden, 1991, p. 138)

And when Mendelian genetics was integrated with molecular biology, "[p]rogress occurred [...] by discovering mechanisms with working entities of different sizes." "Progress in genetics occurred [...] by discovering new mechanisms and integrating them into the temporal series of hereditary mechanisms." (Darden, 1991, pp. 98–99, 120). In another example of the progressiveness of the CF of biology,

> fruitful interfield interactions between molecular biology and biochemistry served to integrate their findings. As work proceeded from both ends of the mechanism, it converged in the middle [...] with the discovery of new entities (three types of RNA and activated amino acids) and their activities. The result was an understanding of the productive continuity of the protein synthesis mechanism from beginning to end. (Darden, 1991, p. 70)

Biology is also epistemically fruitful (e.g., Weber, 2005, pp. 66, 179) and epistemically progressive in other manners (e.g., Weber, 2005, p. 140).

* Specific properties related to practices, methods, and techniques (specifically, properties 88, 92, 94–96, 100, 102–105, 107, 111–113, 115–116, 118, 120, and 154): Biology is rigorous and structured in its research and in its testing practices (Voit, 2019, p. 2; e.g., Darden, 1991, pp. 79, 87, 253). It has clarity as an evaluative criterion (Darden, 1991, p. 299). Biology is painstaking (Weber, 2005, p. 28, see quote below in this section). It systematically gathers information (Weber, 2005, pp. 74, 122–3, 126, 152, 293). Biology uses, and abides by, evidence that it presents or incorporates (Darden, 1991, pp. 29, 32, 49, 52, 74, 91–2, 137–8, 156, 158–62, 164–6, 174, 203, 232, 236–40, 250, 252, 255–6, 259, 261–2, 281–2, 285–6, 288–90, 292–301, 304; Weber, 2005, pp. 62, 95–8, 105, 108, 110–1, 113–7, 123–4, 145, 147, 157, 201–2, 211, 225, 284–5, 292, 301, 303, 314).

> As more empirical evidence is found, and [...] evaluation strategies are successfully employed, the how possibly hypothesized mechanism becomes how plausibly. [...] Plausible hypotheses passing [...] tests nonetheless need to be subjected to direct empirical testing to provide evidence [...]. (Darden, 1991, pp. 292–4)

For example, Watson modified Gamow's idea "that proteins were synthesized directly on the DNA double helix" because "Watson was aware of biochemical evidence that proteins do not form directly on DNA but instead are associated with RNA", and Watson's specific proposal, although it was later disconfirmed, "was consistent with available evidence that RNA was involved in the mechanism" (Darden, 1991, p. 32). This data was and is obtained through carefully controlled experiments (e.g., Darden, 1991, pp. 250, 256, 261, 302; Weber, 2005, pp. 10, 114–26, 129, 131, 134, 146–7, 175, 187, 290–1, 293–4). Biology uses mainly careful scrutiny (Darden, 1991, pp. 30, 261, 290, 296, 302). It contains an adequate view of the scientific method (so, it possesses property 175) and has adopted the general scientific method (described in Sect. 2.2.2.1, and including both observation-driven and hypotheses-driven approaches) as its main strategy for dealing with its cognitive problematics (Sousa, 2016, pp. 110–1, 115; Voit, 2019, p. 1), with evolving, historically correctable procedures (methods and techniques) (Darden, 1991, pp. 138, 156, 159). Its descriptions and explanations can be independently recognised as correct (Darden, 1991, pp. 47, 136). Biology is analytical but also generalises (Darden, 1991, pp. 19–20, 28, 212–6, 240, 248–9, 252, 255, 261–2, 264, 278,

297, 305; Weber, 2005, pp. 6, 25, 28, 32–5, 42–5, 49, 156, 180–2, 186–7, 273). It does planned research and abides by it; and features theoretically-directed investigation, not just trial and error (Darden, 1991, pp. 30, 48–9, 68, 74, 77, 87, 128–9, 138, 182, 258; Weber, 2005, p. 28: "To figure out what kind of system [... (biologists)] are dealing with, what entities these systems contain, and what physical and chemical principles govern their behavior can take years of painstaking research", pp. 172, 174–6, e.g., p. 95, Chapter 4, and pp. 131–6, 141–3, 281). Experimental biology uses (scientific) hypothesis-driven methods (Meyer et al., 2013), such as the hypothetico-deductive method (Sousa, 2016, pp. 110–1). When experimenting or testing, biology checks for variation of variables (Reiser et al., 2013, p. 264; Sousa, 2016, p. 111). It uses control groups (e.g., Darden, 1991, pp. 59, 250, 256, 261). It gets independent confirmation of the alleged facts (Darden, 1991, pp. 135, 166; e.g., Weber, 2005, pp. 164, 207). For example, the field of "[g]enetics emerged between 1900 and 1905 with the independent [re]discovery of Mendel's law by Hugo de Vries and Carl Correns and with the promulgation of Mendel's experimental approach by William Bateson." (Darden, 1991, pp. 135, 155). Biology studies complex biological systems "(such as organisms or ecosystems)" (Weber, 2005, pp. 7, 16, quote: p. 19) as complex systems (Bechtel & Richardson, 1993; Voit, 2019, pp. 6, 9). Systemicity is an "[e]specially important criteri[on] [...]. It is important that [...] new theoretical component[s] be systematically connected with the other theoretical components" (Darden, 1991, pp. 211, 217). Biology is critical (Weber, 2005, pp. 68, 77, 94, 123, 230–1, 240, 242, 257, 312; e.g., Darden, 1991, pp. 161, 278). "[E]xperimental results in biology are subjected to critical scrutiny" (Weber, 2005, p. 123). It revises or checks (Darden, 1991, pp. 29, 30, 47–8, 70, 78, 302–5). Biology is not indifferent to criteria of valid evidence ("how evidence is *produced* in the laboratory [...] must reflect how scientists scrutinize data, experimental arrangements, possible sources of error, and so forth in order to generate reliable inferences to theoretical claims.". In biology "it is not sufficient simply to examine the relationship between theory and [putative] evidence.") (Weber, 2005, pp. 118–9, 130).

* Specific formal properties of propositions, hypotheses, theories, and arguments (specifically, property 174): In biology, "mathematical population genetics and experimental and field studies of populations" have "provided the key knowledge about mechanisms operating at" "the populational level" (Darden, 1991, p. 172). Also, biology uses mathematical modelling, which involves a collection of up-to-date logical and mathematical theories, to sharpen theory formation and to process data. Sometimes, "the insights gained inductively from computational modeling would have been difficult to obtain purely with experimental laboratory methods" (Voit, 2019, pp. 6–9). In order to do this,

> the system encompassing the hypothesis is moved into the realm of mathematics. This move requires two sets of ingredients. One set consists of the simplification and abstraction of the biological system: Any distracting details that seem unrelated to the hypothesis and its context are omitted or represented collectively with other details. [...] The second set of ingredients consists of correspondence rules that translate every biological component or process into the language of mathematics [...]. (Voit, 2019, pp. 7–8)

* Specific properties related to the contents of problems, propositions, hypotheses, and theories (semantical aspects, metaphysical aspects, aboutness, etc.) (specifically, properties 175–176, 179, 182–184, and 188–190): Regarding the possession of the property 175, see the paragraph on specific properties related to practices, methods, and techniques. Biology changes progressively because of the anomalies it encounters (e.g., Darden, 1991, p. 60). It features a high degree of specificity in the theoretical description of the causal mechanisms that its theories propose (Darden, 1991). Although biology in its current state is not as legal as other natural sciences, it does propose explanations with respect to laws[5] (Ruse, 1970; Darden, 1991, pp. 17–20; Schaffner, 1993, pp. 287, 306–7; Sousa, 2016, pp. 111–2, 115). For example,

> [t]he field of *classical Mendelian genetics* emerged in 1900 with its central problem being the explanation of patterns of inheritance of characteristics. The technique used was artificial breeding of organisms with variant characteristics. Empirical regularities were explained by appeal to the two Mendelian laws of segregation and independent assortment, which state regularities in the behavior of hypothetical genes. [...] Mendelian genetics could be analyzed as having two laws as part of the theory of the gene [...] (Darden, 1991, pp. 100, 103)

Biology is empirically adequate within its intended domain of application (Darden, 1991, pp. 20–5, 112, 209, 217, 256–8, 288–9, 300, 306) and enjoys a high degree of theoretical confirmation (e.g., Darden, 1991, pp. 58, 79, 130–1, 137–8, 162, 239). It uses simple explanations (e.g., Weber, 2005, pp. 65–6). It is contiguous with other current sciences, especially physics and chemistry (Darden, 1991, pp. 65, 132, 145–6). The three fields share items in their general outlooks, formal backgrounds, specific backgrounds, funds of knowledge, aims, and methodics; and each member of the domain of one of the fields is a component of a system in the domain of the other (for example, both particles and chemical bonds are components of living things). This contiguity shows that biology is bulk externally consistent with current science.

* Specific properties related to epistemic attitudes and values (axiological properties), individual or social (specifically, properties 195, 201, and 210 (which requires property 51)): Biologists look for regularities, for example, mechanisms (Scriven, 1959, p. 481; Darden, 1991, pp. 15, 20, 100; Weber, 2005, pp. 30, 45, 64, 285). Their research community is a social system composed of persons who continue traditions of inquiry (Darden, 1991, pp. 28, 72, 130; Weber, 2005, pp. 80–7, 168–9). Biology changes progressively because of the anomalies that it encounters (e.g., Darden, 1991, p. 60 and Part II; Weber, 2005, Chapter 3).

---

[5] The existence of distinctly biological universal laws has been challenged, for example by Smart (1963, pp. 50–61), by Beatty's Evolutionary Contingency Thesis (1995), and for all the so-called "special" sciences (i.e., sciences other than fundamental physics) by Earman and Roberts (1999). Ruse (1970) rebuked Smart, and several authors (Carrier, 1995; Waters, 1998; Weber, 1999) have criticized the Evolutionary Contingency Thesis. It is probably too strong a requirement that the "laws" in the property 182 must be universal. If it is too strong a requirement, then proposing explanations with respect to distinctly biological universal laws is not required for biology to possess the property 182.

* Specific epistemic properties of practices, propositions, hypotheses, and theories, and of the evolution of the discipline (specifically, properties 212 and 215): The biology research community evaluates theories, models, and explanations in part by considering their alternatives ("rivals") (Darden, 1991, pp. 215, 272, 289–91, 294, 299–300, 305–6; e.g., p. 304, Weber, 2005, pp. 62, 97, 99–100, 199, 218). Biology changes progressively because of the anomalies that it encounters (e.g., Darden, 1991, p. 60 and Part II; Weber, 2005, Chapter 3). In this and other respects, it is self-corrective (Darden, 1991, pp. 2, 272, e.g., pp. 33, 137).

* Properties of various types related to problem-solving (specifically, properties 40, 191, and 216): Biology solves cognitive problems (Darden, 1991; Bechtel & Richardson, 1993), and has solved theoretical problems of its own. For example, "[i]f DNA is transcribed into ribosomal RNA [(as it was believed in the 1950s)], one would expect the base ratios of the DNA and the ribosomal RNA in a given species to be similar. They were not. [...]. The discovery of [...] messenger RNA is how the ribosomal base anomaly was soon resolved in 1961." (Jacob & Monod, 1961; Darden, 1991, pp. 5, 211–8, e.g., pp. 79 (quote), 224).

17.3 The Properties That Biology (Mostly) Does Not Have

There are concerns about the degrees of precision and unambiguity of the definitions of some of the most important terms in biology (e.g., "life", "species", "gene"). "There are hundreds of working, conventional definitions of life within scientific discourse, but none has been able to achieve a consensus" (Tsokolov, 2009, p. 401). "[W]e still cannot define the term "species" with any degree of confidence and biologists nearly come to blows over what constitutes this taxon" (Brown, 1994, p. 144). "[W]e are still far from having reached unanimity on […] the delimitation of species taxa" (Mayr, 2004, Chapter 10, quote: pp. 171–2) (see also Sober, 2000, Chapter 6; Godfrey-Smith, 2014, pp. 100–7) (lack of property 36).

We have already dealt (in Sect. 17.2) with the concerns about the existence and quantity of distinctly biological universal laws in biology.

Regarding all the properties in Sect. 2.2 not mentioned in this chapter, it is clear that biology mostly has most of them.

17.4 Chapter Conclusions

Biology has at least 35 of the specific properties of science featured in the demarcatory list, including properties of all the types mentioned in the list. There are legitimate concerns about biology lacking the property 36 to a considerable degree. This yields a score of 34 for biology. This provisional tally stands strongly in favour of the scientificity of biology, and is the best explanation for the fact that biology is commonly deemed scientific. This, once again, demarcates in a manner that is equivalent to that of the current international philosophical and scientific consensus regarding the extension of the term "science".

References

Beatty, J. (1995). The evolutionary contingency thesis. In G. Wolters, J. G. Lennox, & P. McLaughlin (Eds.), *Concepts, theories, and rationality in the biological sciences. The second Pittsburgh–Konstanz colloquium in the philosophy of science* (pp. 45–81). University of Pittsburgh Press.

Bechtel, W., & Richardson, R. C. (1993). *Discovering complexity*. Princeton University Press.

Brown, A. C. (1994). Is biology a science? *Transactions of the Royal Society of South Africa, 49*(2), 141–146.

Carrier, M. (1995). Evolutionary change and lawlikeness, beatty on biological generalizations. In G. Wolters, J. G. Lennox, & P. McLaughlin (Eds.), *Concepts, theories, and rationality in the biological sciences. The second Pittsburgh–Konstanz colloquium in the philosophy of science* (pp. 83–97). University of Pittsburgh Press.

Curtis, H., et al. (2014). *Biología* (7th ed.). Editorial Sudamericana.

Darden, L. (1991). *Theory change in science: Strategies from Mendelian genetics*. Oxford University Press.

Earman, J., & Roberts, J. (1999). "Ceteris Paribus", there is no problem of provisos. *Synthese, 118*(3), 439–478.

Godfrey-Smith, P. (2014). *Philosophy of biology*. Princeton University Press.

Jacob, F., & Monod, J. (1961). Genetic regulatory mechanisms in the synthesis of proteins. *Journal of Molecular Biology, 3*, 318–356.

Kitcher, P. (1984). 1953 and all that: A tale of two sciences. *The Philosophical Review, 93*, 335–373.

Lack, C., & Rousseau, J. (2016). *Critical thinking, science, and pseudoscience: Why we can't trust our brains*. Springer.

Mayr, E., & Provine, W. B. (1980). *The evolutionary synthesis: Perspectives on the unification of biology*. Harvard University Press.

Mayr, E. (2004). *What makes biology unique? Considerations on the autonomy of a scientific discipline*. Cambridge University Press.

Meyer, S. C., Nelson, P. A., Moneymaker, J., Minnich, S., & Seelke, R. (2013). *Explore evolution: The arguments for and against neo-Darwinism*. Hill House.

Popper, K. (1978). Natural selection and the emergence of mind. *Dialectica, 32*, 339–355.

Reiser, B. J., Tabak, I., Sandoval, W. A., Smith, B. K., Steinmuller, F., & Leone, A. J. (2013). BGuILE: Strategic and conceptual scaffolds for scientific inquiry in biology classrooms. In S. M. Carver & D. Klahr (Eds.), *Cognition and instruction: Twenty-five years of progress*. Erlbaum.

Ruse, M. E. (1970). Are there laws in biology? *Australasian Journal of Philosophy, 48*(2), 234–246.

Schaffner, K. (1993). *Discovery and explanation in biology and medicine*. University of Chicago Press.

Scriven, M. (1959). Explanation and prediction in evolutionary theory. *Science, 130*, 477–482.

Smart, J. J. C. (1963). *Philosophy and scientific realism*. Routledge and Kegan Paul.

Smocovitis, V. B. (1992). Unifying biology: The evolutionary synthesis and evolutionary biology. *Journal of the History of Biology, 25*(1), 1–65.

Sober, E. (2000). *Philosophy of biology* (2nd ed.). Westview Press.

Sousa, C. (2016). The scientific methods of biology, starting with Charles Darwin. *The American Biology Teacher, 78*(2), 109–117.

Treviranus, G. R. (1802). *Biologie, oder Philosophie der lebenden Natur* (Vol. 1). Röwer. Unattributed trans. quoted in Richards, R. J. (2002), *The Romantic Conception of Life: Science and Philosophy in the Age of Goethe*. Chicago: University of Chicago Press.

Tsokolov, S. A. (2009). Why is the definition of life so elusive? Epistemological considerations. *Astrobiology, 9*(4).
Voit, E. O. (2019). Perspective: Dimensions of the scientific method. *PLoS Computational Biology, 15*(9), e1007279.
Waters, C. K. (1998). Causal regularities in the biological world of contingent distributions. *Biology and Philosophy, 13*, 5–36.
Weber, M. (1999). The aim and structure of ecological theory. *Philosophy of Science, 66*(1), 71–93.
Weber, M. (2005). *Philosophy of experimental biology*. Cambridge University Press.

Chapter 18
The Properties of Psychology

Abstract This chapter provides a brief definition and a description of/introduction to psychology. It then meta-analytically shows that the demarcation criteria and the scoring method proposed in Chap. 2 demarcate psychology as scientific, in agreement with the current philosophical and scientific consensus.

Keywords Psychology · Scientific demarcation · Science · Scientificity

18.1 Introduction

This chapter shows that the demarcation criteria demarcate current academic psychology (CAP) as scientific, in agreement with the current philosophical and scientific consensus. This section provides a working definition of the term "CAP". Section 18.2 lists properties of scientificity that CAP has. Section 18.3 lists properties of scientificity that CAP lacks. As the conclusion of this chapter, Sect. 18.4 argues that the demarcation criteria listed in Sect. 2.2 demarcate CAP as scientific (that current academic psychology has enough of the properties of scientificity, and therefore should be deemed scientific), and that the fact that most CAP is commonly deemed scientific is explained by the fact that it has enough of the properties of science listed in Chap. 2.

Psychology is the study of, and theory on, the varied conscious and unconscious internal phenomena that occur in individual minds of complex animals, including in their relations with others (i.e., social psychology). These phenomena are mental states and processes, including feeling, thought, perception, intelligence, cognition, memory, attention, learning, emotion, moods, intelligence, subjective experiences, motivation, personality—i.e., enduring patterns of behaviour, thought, and emotion—, their development, the behaviour that those minds, or the owners of those minds, perform, the roles of mental states, processes, and functions in those behaviours, and aggression. "Psychology is the [...] *study of behavior and cognitive processes*. As such, psychology describes thinking and behavior and looks at the

D. Fernandez-Beanato, *A Working Scientific Demarcation*, Synthese Library 519, https://doi.org/10.1007/978-3-031-99005-2_18

relationships between them (the what) and tries to explain the causes for them (the why)" (Wittig, 2001, p. 1).

Most CAP is commonly deemed scientific, and a behavioural science (Wittig, 2001, p. 1; see e.g. Barker et al., 2002, pp. 20–4; Gray & Bjorklund, 2018, pp. 48–51, 71: CAP "is a vast and diverse field. Every question about behavior and mental experience that is potentially answerable by scientific means is within its scope"). (There have been, and there can be, non-scientific psychologies, but these, as they are not usually part of CAP, do not concern us in this chapter.) As with all scientific CFs, a psychological field is scientific as long as it possesses enough of the properties of scientificity and is the one that contextually (in a given period and society) has the most of them (in this way, previous developments stop being scientific as soon as a (demonstrably) sufficiently epistemically-progressive integral separated approach develops).

Although psychology has social aspects and subdisciplines (e.g., social psychology) and is sometimes incorrectly classified as a social science, it is eminently about individuals analysed qua individuals, and therefore it is not a social science. Rather, psychology is sui generis in that it is a science that is not a formal, a natural, or a social science (Gaeta et al., 2007).

Nonetheless, this work does not make any assumptions on the advisability of the current fragmentation of behavioural science into discrete disciplines, such as psychology and sociology.

The concern of this chapter is what properties of science psychology has and what properties of science it lacks. This chapter does not provide reasons for believing that CAP is a science other than those that already exist in the literature. I take the properties of science listed in Sect. 2.2, show that CAP possesses many of them, and that this is asserted by the specialists on CAP in the context of discussing what CAP is like. This provides: a) a rational reconstruction (and therefore an explanation) of *why* most CAP is scientific, b) further evidence of the extensional adequacy of this work, c) further evidence of this work's explanatory power, and d) further evidence of the fact that the proposed set of criteria demarcates science appropriately.

This work assesses CAP by using scholarly literature. This chapter, like all other case studies in this work, incorporates analyses performed by several distinguished authors who are specialists in the field and who concluded that CAP possesses or lacks properties that this work, in turn, uses for SD. I rely here on mainstream authoritative literature on CAP and on the philosophy of psychology written by practising psychologists and by philosophers of science. The texts used are key to this investigation because they engage with the properties of science that are featured in CAP, *as* they are featured in CAP. These attributions of properties are mutually commensurable (as part of mainstream traditions in the philosophy of science) and largely uncontroversial. They were arrived at by using different methods and are, in many cases, also independent of our problem, in the sense that the authors' concern when making them was not SD. All of these works are sufficiently up-to-date and standard, not controversial, with respect to the points taken up here. For all the above-mentioned reasons, we can rely on this set of authors for our purposes.

Therefore, this work recognises that CAP has the properties that those scientific or philosophical authorities adjudicated to it. In this way, this work is meta-analytical, in that its analysis uses and incorporates the conclusions of these authors' original analyses.

Nonetheless, explicit claims or arguments about CAP's possession of properties of science are difficult to find in the literature because the possession of these properties by CAP is usually taken for granted, and that is why most of the properties listed in Chap. 2 are not covered in this chapter, although it is clear that CAP possesses most of them.

From now on, this chapter uses the term "psychology" to refer to CAP.

As a scientific CF, psychology should not be confused with parapsychology, which is treated in this work as a clear case of a non-scientific CF (in Chap. 4). Parapsychology was ostracised by the international psychological community after 1905 (Benjamin & Baker, 2012, pp. 3–4).

It is not necessary that all people who claim or are said to be scientific psychologists or scientific psychological researchers are in fact scientific, or that all people who claim to do scientific psychology in fact do it. Some of them are non-scientific, in different ways, some of them in quite extreme ways. For example, some "may not claim to represent, or may even reject the idea of, [the] reality [of our world]." (Barker et al., 2002, p. 2). This

> can be found in the various forms of *constructionism* and *constructivism*, some of which overlap considerably with *postmodernism* (Gergen, 2001; Guba & Lincoln, 1989; Lyotard, 1979/1984; Neimeyer, 1993). These are fairly imprecise terms, but they share a common stance of dispensing with the assumption of an objective reality [...]. Postmodernists are impatient with what they call "grand theory"; instead they present a more multifaceted, fractured world view, some taking the extreme point of view that there are no true and false stories, only different stories. The central problem with such radical constructionist or postmodernist views is that not all constructions are equally accurate, consistent, consensually replicable, or useful. (Barker et al., 2002, p. 12)

From now on, this chapter does not refer to non-scientific so-called "psychology".

It is also not necessary that all scientific psychologists think of what they do as scientific psychology (even if it is so). And it is not necessary that all scientific psychologists always do scientific psychology or live up to its ideals.

In any case, clearly, there is a substantial body of work (by a substantial body of psychologists) having properties that correspond to those of the kind of psychology that is described in this chapter as scientific. It is because this body of psychology possesses those properties that it is deemed scientific here. This chapter establishes that insofar as there is this way of doing psychology, it is scientific (and it is recognised as such), and it is so because it possesses the properties of science listed here. Whether or not this kind of psychology is a large part of psychology, or a large part of what is called "psychology" or of what is called "scientific psychology" in other works, or whether or not what is called "psychology" or "scientific psychology" in other sources has these properties, is not my concern here. This work is also not concerned with what the boundaries of scientific psychology are. Instead, this work is concerned with the characterisation of the kind of psychology that is scientific.

Let us see how psychology stands regarding the possession of our scientificity properties.

18.2 The Properties of Psychology

In psychology, it is understood that "[r]esearch should be conducted so that the collection, analysis, and interpretation of behavioral information are done with maximum *objectivity*" (Wittig, 2001, p. 15) (property 2).

Psychology uses statistical reasoning, a form of correlation reasoning/thinking that "applies statistical methods to find patterns": in psychology "[r]esearchers use probability statistics to decide whether effects are 'significant'" (Wittig, 2001, pp. 15, 19–25; Dienes, 2008, pp. 55–156; Gray & Bjorklund, 2018, pp. 109, 121–4, 127, 129, 132–40; quotes: Thagard, 2011, p. 27; Coolican, 1996, p. 3) (property 23).

Psychology generates (or has generated) hypotheses and theories through deductive and non-deductive processes (cf. Coolican, 1996, pp. 7–9; Barker et al., 2002, pp. 14–7) (property 31).

Psychology presupposes philosophical background (underlying) assumptions (Barker et al., 2002, p. 6) (property 37).

Psychology offers accounts such as descriptions, hypotheses, theories, and models (Dienes, 2008, pp. 14–7, 23–4, 26, 28, 34, 46; Gray & Bjorklund, 2018, pp. 113, 117, 124–7) (property 42).

Scientific psychology is falsifiable (cf. Dienes, 2008, pp. 1–32): "Psychological researchers generally follow a scientific approach. This involves the logic of testing hypotheses produced from falsifiable theories." (Coolican, 1996, p. 3) (properties 52 and 54).

In psychology, "an experimenter manipulates a variable to be studied, chooses the response to be measured, and controls extraneous influences that might inappropriately affect the results of an experiment" (Wittig, 2001, pp. 15–8, quote: p. 15). "Research has to be carefully planned with attention to design, variables, samples and subsequent data analysis. If all these areas are not fully planned, results may be ambiguous or useless." Although not all researchers adhere to these methods, "a majority of psychologists would agree that research should be scientific, and at the very least that it should be objective, controlled and checkable" (quote: Coolican, 1996, pp. 3–4, 17; Gray & Bjorklund, 2018, pp. 114–5) (properties 51, 55, 96, 102–103, 111–112, 147–148, and 154).

Psychological research results that are not repeatable/replicable may be considered suspect and not widely accepted (Wittig, 2001, p. 16). Results in psychology have been replicated (Gray & Bjorklund, 2018, pp. 153–6) (properties 56–57).

Psychology is generally rigorous and requires clarity (properties 88 and 92):

> [T]here are general principles common to rigorous research within whatever paradigm [in psychology], and [...] it is possible to lay out an overall framework which organizes different approaches to research and clarifies the ways in which they can complement one

> another. Learning to do psychological research is partly a process of learning disciplined enquiry according to these principles within this general framework.
>
> [...]
>
> Research [in psychology] is demanding: it does require clear and rigorous thought [...] (Barker et al., 2002, pp. 3–4)

Psychology, especially experimental psychology, uses the general scientific method (Wittig, 2001, p. 7) (properties 103–104).

Psychology uses control groups (Wittig, 2001, p. 17; Barker et al., 2002, pp. 4, 23) (property 113).

Psychology is critical: psychological research involves critical thinking (Barker et al., 2002, p. 5) (property 118).

Psychologists try and eliminate biases by using finely honed and systematic safeguards (e.g., experimental and sampling techniques) to compensate for cognitive biases (Wittig, 2001, pp. 15, 18–9; Gray & Bjorklund, 2018, pp. 141–4, 146) (properties 51 and 147–148).

Psychology is contiguous with, and connected to, all other types of sciences (Gray & Bjorklund, 2018, pp. 86–8) (property 189).

18.3 The Properties That Psychology (Mostly) Does Not Have

No property of science is indicated in the literature as clearly lacking in scientific psychology.

18.4 Chapter Conclusions

Psychology has at least 14 of the specific properties of scientificity featured in the demarcatory list, while not clearly lacking any property. This yields a score of 14 for psychology. This (possibly provisional) tally stands strongly in favour of the scientificity of psychology, and is the best explanation for the fact that (most) psychology is commonly deemed scientific. This, once again, demarcates in a manner that is equivalent to that of the current international philosophical and scientific consensus regarding the extension of the term "science".

References

Barker, C., Pistrang, N., & Elliott, R. (2002). *Research methods in clinical psychology: An introduction for students and practitioners* (2nd ed.). Wiley.

Benjamin, L. T., Jr., & Baker, D. B. (2012). The internationalization of psychology: A history. In D. B. Baker (Ed.), *Oxford handbook of the history of psychology*. Oxford University Press.

Coolican, H. (1996). *Research methods and statistics in psychology* (2nd ed.). Hodder & Stoughton.
Dienes, Z. (2008). *Understanding psychology as a science: An introduction to scientific and statistical inference*. Palgrave Macmillan.
Gaeta, R., Gentile, N., & Lucero, S. (2007). *Aspectos críticos de las ciencias sociales*. EUDEBA.
Gergen, K. J. (2001). Psychological science in a postmodern context. *American Psychologist, 56*, 803–813.
Gray, P. O., & Bjorklund, D. F. (2018). *Psychology* (8th ed.). Worth Publishers.
Guba, E. G., & Lincoln, Y. S. (1989). *Fourth generation evaluation*. Sage.
Lyotard, J.-F. (1979). *La condition postmoderne: Rapport sur le savoir* (Trans. 1984 as *The Post-Modern Condition: A Report on Knowledge*). Manchester University Press.
Neimeyer, R. A. (1993). An appraisal of constructivist psychotherapies. *Journal of Consulting and Clinical Psychology, 61*, 221–234.
Thagard, P. (2011). Evolution, creation, and the philosophy of science. In R. S. Taylor & M. Ferrari (Eds.), *Epistemology and science education: Understanding the evolution vs. intelligent design controversy* (pp. 20–37). Routledge.
Wittig, A. F. (2001). *Introduction to psychology*. McGraw-Hill.

Chapter 19
The Properties of Sociology

Abstract This chapter provides a brief definition and a description of/introduction to sociology. It then meta-analytically shows that the demarcation criteria and the scoring method proposed in Chap. 2 demarcate sociology as scientific, in agreement with the current philosophical and scientific consensus.

Keywords Sociology · Scientific demarcation · Science · Social sciences · Scientificity

19.1 Introduction

This chapter shows that the demarcation criteria demarcate most of mainstream current academic sociology (CAS) as scientific (as a social science), in agreement with the current philosophical and scientific consensus. This section provides a working definition of the term "CAS". Section 19.2 lists properties of scientificity that CAS has. Section 19.3 lists properties of scientificity that CAS lacks. As the conclusion of this chapter, Sect. 19.4 argues that the demarcation criteria listed in Sect. 2.2 demarcate CAS as scientific (that CAS has enough of the properties of scientificity, and therefore should be deemed scientific), and that the fact that most CAS is commonly deemed scientific is explained by the fact that it has enough of the properties of science listed in Chap. 2.

CAS is the CF that is about human societies, aggregated human social behaviour and its patterns, relationships, social interaction, social structures, culture, social order, social development, and social change.

Most CAS is scientific; it is a social science, and therefore an empirical science (Hempel, 1962; Duneier et al., 2007, p. 24; Ashley & Orenstein, 2005; Ferrante, 2010, p. 45; Thompson & Hickey, 2011, pp. 2, 7, 34; Schaefer, 2012, p. 5; Ferris & Stein, 2016, p. 9; Browne, 2020, pp. 8–9, 43; van Tubergen, 2020, pp. 3, 35, 75). It is "the scientific study of social life" (van Tubergen, 2020, p. 3) and social phenomena (ibid., pp. 9, 35). The concern of this chapter is what properties of science it has and what properties of science it lacks. This chapter does not provide reasons for

D. Fernandez-Beanato, *A Working Scientific Demarcation*, Synthese Library 519, https://doi.org/10.1007/978-3-031-99005-2_19

believing that CAS is a science other than those that already exist in the literature. I take the properties of science listed in Sect. 2.2, and show that CAS possesses many of them, according to specialists on CAS in the context of discussing what CAS is like. This provides: a) a rational reconstruction (and therefore an explanation) of *why* most CAS is scientific, b) further evidence of the extensional adequacy of this work, c) further evidence of this work's explanatory power, and d) further evidence of the fact that the proposed set of criteria demarcates science appropriately.

This assessment of CAS uses scholarly literature. This chapter, like all other case studies in this work, incorporates analyses performed by several distinguished authors who are specialists in the field and who concluded that CAS possesses or lacks properties that this work, in turn, uses for SD. I rely here on mainstream authoritative literature on CAS, including many introductory textbooks. Most of the texts used here are issued by prestigious international academic publishers, are extensively utilised in higher education, and have had numerous editions (e.g., Ashley & Orenstein, 2005; Duneier et al., 2007; Macionis & Plummer, 2008; Tischler, 2010; Thompson & Hickey, 2011; Schaefer, 2012; Giddens et al., 2013), or are some of the latest textbooks in the sociological field (e.g., Browne, 2020; van Tubergen, 2020). These texts are key to this investigation because they engage with the properties of science that are featured in CAS, *as* they are featured in CAS. These attributions of properties are mutually commensurable. They were arrived at by using different methods and are, in many cases, also independent of our problem, in the sense that the authors' concern when making them was not SD. This chapter also analyses sociology based on the norms, rules, and recommendations for the practice of the discipline put forward in these introductory textbooks. All of these works are sufficiently up-to-date and standard, not controversial, with respect to the points taken up here. For all the above-mentioned reasons, we can rely on this set of authors for our purposes. Therefore, this work recognises that CAS has the properties that those scientific authorities adjudicated to it. In this way, this work is meta-analytical, in that its analysis uses and incorporates the conclusions of these authors' original analyses.

From now on, this chapter uses the term "sociology" to refer to scientific CAS.

The fact that this chapter defines "sociology" the way it does implies that the chapter is not about other things that could conceivably be called "sociology". There might be work in sociology that lacks some of the properties adjudicated in this chapter to sociology. Not all things called "sociology" are necessarily scientific, and not all persons called "sociologists" or employed in academic Departments of Sociology are necessarily scientific sociologists. And it is not necessary that all sociologists who claim or are said to be scientific sociologists are in fact scientific sociologists, or that all sociologists who claim to do scientific sociology in fact do it. It is also not necessary that all scientific sociologists think of what they do as scientific (even if it is so). And it is not necessary that all scientific sociologists always do scientific sociology or live up to its ideals.

The authoritative authors referenced in this chapter agree with me that all of sociology that is worth the name is broadly scientific. But even without committing to this position we can clearly see that a large body of the work of sociology (made by a substantial body of sociologists) is scientific (in line with the definitions given in Chap. 1), and recognise that a large body of sociologists say so. This body of work is of the kind of sociology that this chapter is concerned with. It is deemed scientific by the experts used in this chapter and by me because it possesses the properties of science listed in this chapter, i.e., the properties of the kind of sociology that is described in this chapter as scientific. This chapter argues that insofar as there is this way of doing sociology, it is scientific (and it is recognised as such), and it is so because it possesses the properties of science listed here. Many textbooks on sociology equate scientific sociology, or the scientific approach in sociology, with "positivism". This is not how the term "scientific sociology" is used in this chapter, as it is clear that also subjective/empathic understanding (*Verstehen*) and interpretation have a legitimate place in scientific sociology, for example on the lines of more recent multilevel sociological models (see e.g., van Tubergen, 2020, pp. 123–32). Whether or not scientific sociology is a large part of what is called "sociology" in other works, or whether or not what is called "sociology" in other sources has these properties, is not my concern here. Instead, this work is concerned with the characterisation of the kind of sociology that is scientific.

Let us see how sociology stands regarding the possession of our scientificity properties.

19.2 The Properties of Sociology

Sociology uses logical argumentation and reliable data (it is based on all the evidence and abides by it) to evaluate its theories (Duneier et al., 2007, pp. 24, 34; Thompson & Hickey, 2011, pp. 6, 30, 51; Schaefer, 2012, pp. 31–2, 34; Browne, 2020, pp. 10, 68, 80; van Tubergen, 2020, pp. 74, 76) (properties 1, 29, 49, 66, 96, and 174). It "involves systematic methods of empirical investigation" (Duneier et al., 2007, p. 24), employs various and diverse "methods of empirical investigation and critical analysis" and is sceptical (Ashley & Orenstein, 2005, pp. 3–5; Thompson & Hickey, 2011, pp. 6–7, 23, 31: "thinking sociologically is a form of **critical thinking**"; Browne, 2020, p. 2; van Tubergen, 2020, p. 76) (properties 42, 51, 96, 107, and 118), including both deductive and non-deductive means/reasoning, and it features confirmations and disconfirmations (Duneier et al., 2007, pp. 34–5; Macionis & Plummer, 2008, pp. 74–5; Thompson & Hickey, 2011, pp. 31, 34, 51; van Tubergen, 2020, pp. 47, 91) (properties 7 and 31).

Sociology is systematic (Thompson & Hickey, 2011, pp. 2, 6–7, 30–31, 34; Schaefer, 2012, p. 31): sociological knowledge is obtained by methods based on systematic observation: "sociology involves the organized, systematic study of phenomena (in this case, human behavior) in order to enhance understanding" (Schaefer, 2012, p. 5). Sociology carefully observes, records observations, collects data, and plans and designs research in a highly systematic manner; its studies and knowledge are systematic (Green & Johns, 1966, pp. 6, 14–8; Macionis & Plummer, 2008, p. 76; Ferrante, 2010, pp. 33–39, 45; Tischler, 2010, pp. 33–5, 42; Thompson &

Hickey, 2011, pp. 23, 34, 40; Schaefer, 2012, pp. 5, 31, 35–6; Ferris & Stein, 2016, p. 9; Browne, 2020, pp. 2, 10, 78; van Tubergen, 2020, p. 75). It has a strong focus "on the careful assessment of evidence" (Giddens et al., 2013, p. 5) (properties 95–96, 102, 111, and 178). It carefully analyses its data (Ferrante, 2010, pp. 43, 45; Tischler, 2010, pp. 39–42) (properties 102 and 107). It strives to avoid biases (Tischler, 2010, pp. 38–9; Browne, 2020, pp. 99–101, 110; van Tubergen, 2020, p. 75) (property 147).

Sociology tries to maintain open-mindedness, "readiness to accept the results of [(...) its] investigations, whatever they may be". Sociology is concerned with the truth (Browne, 2020, pp. 70, 80, 108; van Tubergen, 2020, pp. 46–8, 70, 106) (properties 3, 39, 69, 194, and 196), strives for objectivity, is objective, and uses scientific methods and procedures to ensure as much objectivity as possible and to hone safeguards against bias, including approaching knowledge in a self-correcting manner (Harvey & MacDonald, 1993, p. 123; Ashley & Orenstein, 2005, p. 35; Macionis & Plummer, 2008, pp. 54–5, 60, 76–7; Thompson & Hickey, 2011, pp. 6–7, 30–1; Schaefer, 2012, pp. 5, 31; Browne, 2020, p. 10) (properties 2, 13, 51–52, 54, 69, 75, 79, 104, 148, 199, and 217). Sociology uses control groups (Thompson & Hickey, 2011, p. 40) (property 113). And sociologists are usually prepared to submit evidence to be scrutinised by their peers (Browne, 2020, p. 10) (properties 64 and 100).

Sociology clarifies/defines concepts and is clear (Macionis & Plummer, 2008, pp. 59, 73; Thompson & Hickey, 2011, pp. 6, 32–3, 51; Ferris & Stein, 2016, p. 9), for example in its definitions of problems (Thompson & Hickey, 2011, p. 6; Schaefer, 2012, pp. 31–2) (property 92). Sociology is precise (Duneier et al., 2007, pp. 34–5; van Tubergen, 2020, pp. 21–4, 35), including definitional and theoretical explicitness and precision (van Tubergen, 2020, pp. 49–52, 59, 60, 70), collects precise information (Schaefer, 2012, p. 5) (property 92), uses stringent standards and is rigorous (Macionis & Plummer, 2008, p. 66; Schaefer, 2012, p. 31; cf. Thompson & Hickey, 2011, p. 36), for example when testing (van Tubergen, 2020, pp. 75, 106) (property 88). Sociology is comprehensive and self-coherent/self-consistent (Schaefer, 2012, p. 31; Ferris & Stein, 2016, p. 9; van Tubergen, 2020, p. 46), for example in its explanations (Schaefer, 2012, p. 8) (properties 29, 42, and 46). It uses the general scientific method as its main strategy for dealing with its cognitive problematics (Green & Johns, 1966, pp. 1, 6, 30–7; Vander Zanden, 1988, pp. 43–5; Ferrante, 2010, p. 45; Thompson & Hickey, 2011, pp. 34–5, 51; Schaefer, 2012, p. 31) (properties 51 and 104).

Sociology checks its results, sources, and evidence (van Tubergen, 2020, p. 76) (properties 19 and 159). It presents empirical evidence in support of its findings, propositions, results, hypotheses, and theories (Green & Johns, 1966, p. 4; Duneier et al., 2007, p. 24) (property 100). These are checkable, testable (Green & Johns, 1966, pp. 4, 6, 13–4; Thompson & Hickey, 2011, pp. 31, 34, 51; Browne, 2020, p. 10) (property 51), in-principle fallible (Thompson & Hickey, 2011, p. 6) and in-principle falsifiable (sociology looks for falsifiability for its propositions (van Tubergen, 2020, pp. 49, 62)) (Harvey & MacDonald, 1993, pp. 134–5; Thompson & Hickey, 2011, pp. 34, 36) (property 54), open to inspection and criticism (Browne, 2020, p. 10), confirmable/verifiable (Thompson & Hickey, 2011, p. 34), confirmed/

verified (property 184), and re-testable (Browne, 2020, p. 10). Sociology's findings and results are repeatable/replicable, and repeated/replicated, which is an important criterion for their acceptance in sociology (Macionis & Plummer, 2008, pp. 55, 60, 74–5, 77; Tischler, 2010, p. 42; Thompson & Hickey, 2011, pp. 30–1, 33; Browne, 2020, pp. 70–1, 80–1, 108, 110; van Tubergen, 2020, pp. 76, 80–2, 103–7) (properties 56–57 and 118). Sociology strives for empirical success and adequacy (van Tubergen, 2020, pp. 47–8, 52). Sociology uses mathematical theories and uses, and recognises the epistemic value of, statistics (Green & Johns, 1966, pp. 16–7, 22–7; Macionis & Plummer, 2008, pp. 52, 60, 74–5; Tischler, 2010, p. 42; Thompson & Hickey, 2011, pp. 30, 33–5, 38, 43–4, 51; Schaefer, 2012, pp. 22, 33–6, 39, 41, 43–4, 47; Browne, 2020, pp. 61–9, 110) (property 23). It mainly uses representative samples and randomisation (Tischler, 2010, pp. 35–9; Thompson & Hickey, 2011, p. 41; Browne, 2020, pp. 83–7, 110) (properties 32 and 97) (and consequently its propositions are representative (Browne, 2020, pp. 80, 108)).

Sociology identifies and describes social phenomena/processes, and looks for, finds, and describes correlations or "meaningful patterns" in them (Harvey & MacDonald, 1993, p. 134; Duneier et al., 2007, pp. 36–8; Macionis & Plummer, 2008, pp. 61–2, 77; Ferrante, 2010, p. 43; Thompson & Hickey, 2011, p. 34; Schaefer, 2012, pp. 8, 32; Browne, 2020, p. 2; van Tubergen, 2020, pp. 3, 16–8, 35, 73, 75–6, 106) (properties 23 and 42). It strives for knowledge of the causal mechanisms of the social reality (Harvey & MacDonald, 1993, pp. 7–8; Duneier et al., 2007, pp. 36–8; Macionis & Plummer, 2008, pp. 57, 61–2, 77; Thompson & Hickey, 2011, p. 33) (properties 40 and 70). It is "sophisticated in the definition, identification, and measurement of what are taken to be the causes of social action" (Bruce, 2018, p. 18) (properties 23, 36, 48, 70, 92, and 179). Sociology explains social patterns, processes, and phenomena (Ferrante, 2010, pp. 27–32; Bruce, 2018, pp. 19–21, 27; Browne, 2020, p. 2; van Tubergen, 2020, pp. 3, 16–8, 35, 73, 75, 110, 123–32) (properties 42 and 82) by using methods of scientific explanation (Thompson & Hickey, 2011, pp. 31) (properties 42, 46, and 105). It deals with causation, looks for the underlying causes of processes and events (Browne, 2020, p. 81; van Tubergen, 2020, pp. 3, 16–8, 35, 37, 53, 62–4, 70, 87–8, 107, 123–32), offers causal explanations (van Tubergen, 2020, pp. 114, 124, 126–7), and identifies systematic causes (Thompson & Hickey, 2011, p. 30; Schaefer, 2012, pp. 8, 32): "Sociologists seek to identify the *causal* link between variables; the suspected causal link is generally described in the hypothesis (Neuman, 2009)" (Schaefer, 2012, p. 32). Sociologists think in terms of variables and use variables (van Tubergen, 2020, pp. 77–8, 123–33). Sociology controls for other factors (variables) and conditions that could intervene in the phenomena studied (Thompson & Hickey, 2011, p. 33; Schaefer, 2012, p. 35; Browne, 2020, pp. 80–1; van Tubergen, 2020, pp. 58–62) (property 112).

Sociology is general (Browne, 2020, p. 71). While ascribing importance to, and striving for, simplicity (van Tubergen, 2020, pp. 52, 60) (property 188), sociology brings "theoretical order" to phenomena "of social life" (Green & Johns, 1966, p. 2; Denzin, 1970, p. 5; Thompson & Hickey, 2011, pp. 31) (cf. property 47) by putting forward accounts, such as hypotheses (including causal hypotheses), models, and scientific theories (including causal theories) (Duneier et al., 2007, p. 24; Macionis & Plummer, 2008, pp. 74–5; Ferrante, 2010, pp. 27–8, 39, 41; Tischler, 2010, pp. 32–3, 42; Thompson & Hickey, 2011, pp. 31, 34–5; Schaefer, 2012, pp. 7, 32–5; van Tubergen, 2020, pp. 37–8, 41–52, 70–1) which are systematic (van Tubergen, 2020, p. 126) and general (Green & Johns, 1966, p. 3; Thompson & Hickey, 2011, p. 4; Schaefer, 2012, p. 8). Sociology's conclusions are generalisable, and it has a wide scope (Ferrante, 2010, p. 38; Browne, 2020, pp. 80, 108, 110; van Tubergen, 2020, pp. 50–2, 70, 82–6, 106–7) (properties 42 and 46–47).

Sociology predicts (i.e., it features predictive power) and it is fecund/fruitful with respect to predictions and to successful predictions (it features predictive success) (Thompson & Hickey, 2011, p. 31; Schaefer, 2012, p. 8; van Tubergen, 2020, pp. 21, 45–6, 52, 70, 75) (properties 45 and 82). It is epistemically progressive in that it leads to know newly-known facts: "good research frequently leads to the discovery of unanticipated information" (Tischler, 2010, p. 42; Thompson & Hickey, 2011, p. 35) (property 84):

> In sociology, [...] [a]n effective theory may have both explanatory and predictive power. That is, it can help us see the relationships among seemingly isolated phenomena, as well as to understand how one type of change in an environment leads to other changes. (Schaefer, 2012, p. 8)

Sociology is connected and contiguous with the rest of the sciences (properties 59 and 189):

> Sociology, which once insisted that all human behavior is learned, is today more cognizant of the important relationships among biology, chemistry, physiology, and social behavior. This acknowledgment, along with a vast amount of research linking genetics, diet, chemical imbalances, and other variables to human social behavior, has given rise to the creation of [...] a subfield of sociology that explores relationships between biological and environmental influences on humans. Today, most sociologists concede that the only way to understand the limitations of existing knowledge is through more interdisciplinary training and research. (Thompson & Hickey, 2011, p. 31)

Sociology and the understanding it provides are useful, for example, for the design and implementation of social policies (Browne, 2020, pp. 2, 41–3; van Tubergen, 2020, pp. 3, 16–9, 35, 46–52, 70, 73) (property 50).

19.3 The Properties That Sociology (Mostly) Does Not Have

Sociology does not clearly lack any of the properties of science listed in Chap. 2. Regarding all the properties in Sect. 2.2 not mentioned in this chapter, it is clear that sociology mostly has most of them.

19.4 Chapter Conclusions

Sociology has at least 27 of the specific properties of scientificity featured in the demarcatory list, while not clearly lacking any property. This yields a score of 27 for sociology. This (possibly provisional) tally stands strongly in favour of the scientificity of sociology, and is the best explanation for the fact that (most) sociology is commonly deemed scientific (as a social science). This, once again, demarcates in a manner that is equivalent to that of the current international philosophical and scientific consensus regarding the extension of the term "science".

References

Ashley, D., & Orenstein, D. M. (2005). *Sociological theory: Classical statements* (6th ed.). Pearson Education.
Browne, K. (2020). *An introduction to sociology* (5th ed.). Polity Press.
Bruce, S. (2018). *Sociology: A very short introduction*. Oxford University Press.
Denzin, N. K. (1970). *The research act in sociology: A theoretical introduction to research methods*. Butterworth.
Duneier, M., Appelbaum, R., & Giddens, A. (2007). *Introduction to sociology* (6th ed.). W. W. Norton.
Ferrante, J. (2010). *Seeing sociology: An introduction*. Wadsworth.
Ferris, K., & Stein, J. (2016). *The real world: An introduction to sociology* (5th ed.). W. W. Norton.
Giddens, A., Duneier, M., Appelbaum, R., & Carr, D. (2013). *Introduction to sociology* (9th ed.). W. W. Norton.
Green, B. S. R., & Johns, E. A. (1966). *An introduction to sociology*. Pergamon Press.
Harvey, L., & MacDonald, M. (1993). *Doing sociology: A practical introduction*. The MacMillan Press.
Hempel, C. G. (1962). Frontiers of science and philosophy. In R. G. Colodny (Ed.), *Explanation in science and in history* (pp. 9–19). University of Pittsburgh Press.
Macionis, J. J., & Plummer, K. (2008). *Sociology: A global introduction* (4th ed.). Pearson Prentice Hall.
Neuman, L. W. (2009). *Understanding research*. Allyn and Bacon.
Schaefer, R. T. (2012). *Sociology: A brief introduction* (10th ed.). McGraw-Hill.
Thompson, W. E., & Hickey, J. V. (2011). *Society in focus: An introduction to sociology* (7th ed.). Prentice Hall.
Tischler, H. L. (2010). *Introduction to sociology* (10th ed.). Wadsworth.
van Tubergen, F. (2020). *Introduction to sociology*. Routledge.
Vander Zanden, J. W. (1988). *The social experience: An introduction to sociology*. Random House.

Part III
A Working Demarcation

Chapter 20
Concluding Remarks: Theoretical Import and Practical Applications

Abstract This chapter concludes that the proposed method demarcates accurately in every case studied. The chapter then describes the profound theoretical implications and the substantial practical applicability of this novel development in the philosophy of science.

Keywords Science · Scientific demarcation · Non-science · Pseudoscience · Scientificity

Recall Kuhn's hitherto unsolved problem of finding a set of "[d]isciplinary characteristics [which] permit (… [the]) identification [of scientific units of analysis] as […] science", while at the same time locating every science "close to other scientific disciplines and at a distance from disciplines other than science." (2000, p. 214). This work has solved that problem by finding such a set of properties. The method to achieve this has been the following.

In order to develop the multicriterial approach for the demarcation of science, this work started by compiling a list of 220 low-level properties of science that have been deemed key in the literature on SD (in Chap. 2). Then, I tested that list's demarcatory determinations via many case studies of clear cases of most of the main types of CFs: Chapters 3–19 each consider a particular field, including a wide range of clear cases of particular sciences and of non-sciences of different subtypes (natural sciences, social sciences, pseudosciences, non-pseudoscientific non-sciences, etc.), determining whether the field has the properties. This work's method has been based on performing a reasonably broad comparison of many fields of different types, which together are representative of the whole spectrum of fields of human activity claiming epistemic import. This defines a whole new research programme: thousands of fields or subfields could be analysed along these lines, not only current fields but also historical ones, and not only clear cases of sciences and of non-sciences, but also contested cases.

Parts of this chapter were previously published in Fernandez-Beanato (2020b) and Fernandez-Beanato (2021).

D. Fernandez-Beanato, *A Working Scientific Demarcation*, Synthese Library 519, https://doi.org/10.1007/978-3-031-99005-2_20

Some commentators have remarked that this work is largely descriptive. Although this work presents an original proposal in the form of an explication, or a rational reconstruction, that had not been advanced before and that is successful in solving a long-standing important problem in the sense that it works as intended, it is true that the presentation of the proposal in this work features much description. But this description is itself philosophical, not only because it adequately serves the philosophical goal of defining "science", but also because a description is itself philosophical if it employs adequate philosophical methods, and this description of science does. Furthermore, because this is a broadly scientific (analytical philosophical) account of science, it is metascientific in the philosophical sense.

In order to determine whether my multicriterial approach was viable, I meta-analytically checked a) whether it rationally reconstructs and tracks the widespread academic consensus about the extension of the term "science", b) whether it yields an ordering demarcating the set of the clear cases of science from the set of the clear cases of non-science, and, c) whether this order includes a noticeable differentiating gap between those two sets.

This work's 17 case studies used the 220 properties featured on the list in Chap. 2 as SD criteria, subjecting them to be empirically found in consensually clear (agreed-upon) cases of sciences. And I did find them mostly there. The status of these 220 properties as collectively demarcatory of science was substantiated via these case studies, all of which confirmed this work's main thesis: that assessments made solely on the basis of the possession or lack of the relevant properties on the list and the application of the scoring system proposed in Sect. 2.3 determine, in each case, an SD of fields that corresponds to the current international philosophical and scientific consensus regarding the extension of the term "science" in the broad sense.[1] We found that this procedure demarcates the extension of the term "science", at least in all this wide spectrum of cases, exactly as the vast majority of philosophers and scientists would intuitively demarcate it.

This work has assumed at least two large idealisations/simplifications, namely: a) that the possession of a property by a unit of analysis is all or nothing, and b) that no property is more important than any other, including for every unit of analysis and in every stage of that unit's development (i.e., all properties have equal value). Both of these assumptions are implausible. These assumptions were made because the challenge faced was to create a method for tracking scientificity and non-scientificity, or for recovering the existing consensual extensional demarcation of the term "science", by using a crude and simplified model and a blunt scoring method. Some models are epistemically useful even if they do not precisely correspond to reality, and get qualitative features right without being completely accurate (e.g., a physical model ignoring friction). Therefore, if I could make my approach

[1] Even if it turns out that the same demarcatory results can be obtained by using a proper subset of the set of 220 properties, the entire list, besides giving a more complete and more accurate image of science, might still be necessary to assess any of the thousands of possible units of analysis not covered in this work. In any case, it seems clear that all the properties listed in Sect. 2.2 are properties of SBS as it is conceived of in Chap. 1.

work as a first approximation, we could always argue about how to refine it. And, my model works. This reconstruction of the existing intension of the term "science" is a solution to the main problem of SD mentioned in Sect. 1.1. It is a solution because it works, in the sense that it demarcates as we would like an SD to demarcate (i.e., science as science and non-science as non-science).

Success was not guaranteed from the beginning: things could have gone wrong, as my entire approach was and is in-principle falsifiable. What would have counted as a failure, and what would have counted as success for my approach? The goal was to track scientificity and non-scientificity, or to reconstruct the existing consensual extensional demarcation of science, so the weakest kind of success would be that most sciences analysed were ranked above most non-sciences analysed. This I have achieved. A further level of success would be achieved if every science analysed were ranked above every non-science analysed, which I have also achieved. An even further level of success would be achieved if a wide demarcatory gap between the set of sciences and the set of non-sciences were to be found. This has also been achieved.

As part of the project proposed by Mahner (2007, pp. 521–22; 2013, pp. 40, 42), we started this work looking for a demarcatory boundary or "cut": the minimum score (measured in total aggregated amount of properties featured) that a theory, CF, practice, etc. would have to feature in order for it to be legitimately deemed scientific.

The following table (Table 20.1) summarises the main results of our study:

The results are very clear. It is readily noticeable that every clear case of a non-science presents a negatively-valued score, whilst every clear case of an SBS presents a positively-valued score. The highest-valued clear case of non-science is

Table 20.1 Main results of our study (total score of each field)

Field	Total score
Genethliac astrology	−40
Parapsychology	−10
Creation "science"	−18
Homeopathy	−28
Feng Shui	−30
Common sense	−36
Art	−60
Religion	−66
Analytic philosophy	12
Human history	13
The formal sciences	16
Physics	29
Chemistry	17
Astronomy	13
Biology	34
Psychology	14
Sociology	27

parapsychology, with a score of -10, while the lowest-valued clear case of SBS is analytic philosophy, with a score of 12. We can observe that these results rationally reconstruct and track the widespread academic consensus about the extension of the term "science"; that they yield an ordering demarcating the set of the clear cases of science from the set of the clear cases of non-science; and that this order includes a noticeable, wide differentiating gap in values between scientificity and non-scientificity. This gap is 22 points wide.

All the results in this work were accomplished without a need for establishing the "weight" (degree of importance) of each property of science relative to each other's, without considering non-binary degrees of possession of each property, and without computing properties differentially for CFs of different subtypes or in different stages of their respective developments.

The only ordering that I have sought here has been a demarcatory ordering of the set of the sciences and the set of the non-sciences. This has been achieved. But I could also have aspired to obtain an ordered ranking of the sciences reflecting some common ideas about the degrees of scientificity of the different sciences (e.g., many philosophers believe that physics is more scientific than biology). For some of these philosophers, the results in Table 20.1 may seem anomalous; for example, in that the score of biology is greater than that of physics.

Even if these philosophers were right about that (i.e., if the ordering of the sciences that can be inferred from Table 20.1 were wrong), there would be several potential reasons for this anomaly. It might be an artefact of the literature, because the possession of properties of science by some of the more clearly scientific sciences is usually taken for granted in the literature, or because some sciences have to defend their status as sciences more than others (e.g., the scientificity of physics is largely never contested, but there have been doubts about large or important parts of biology, history, sociology, etc.), or are more complex or broad in some ways, involving a need for additional criteria (e.g., biology involves matters of functionality that are absent in physics), or have more rival or challenger fields than others (e.g., scientific biology is challenged by creation "science", but there is no equivalent challenger for physics), and consequently the literature on those fields is more focused on SD and there is more emphasis on what makes the field scientific, which would be picked up by my meta-analyses. Simplicity and the absence of rival fields can be reasons why the literature takes for granted the scientificity of, and the possession of properties of science by, some sciences (e.g., physics) more than it takes for granted the scientificity of, and the possession of scientificity properties by, other sciences (e.g., biology). An artefact of the literature could also be due to the fact that the philosophy of some scientific fields is less developed than that of others (e.g., there is more published philosophy of biology than published philosophy of astronomy or of chemistry). Additional potential reasons for the arguably counter-intuitive or anomalous ranking of the sciences that I have obtained are that in some cases (e.g., philosophy, history, sociology) I have only assessed the parts of the CFs that are more scientific; that there might exist problems of applicability of some of the properties to some of the sciences; and the two simplifications mentioned above (that the possession of a property by a unit of analysis is counted as all or nothing, and that no property is taken to be more important than any other). So this apparently anomalous ordering is not a final result of my model and my method; both could be refined.

But, in any case, my results at this stage are intended to be interpreted only binarily: considering the goals of this work, an incorrect rank ordering is not a problem given that every science is located a long way away from every non-science. Philosophers who believe that there are anomalies in the rank ordering of degrees of scientificity that can be inferred from Table 20.1 need not take this ordering very seriously, given the several above-mentioned potential reasons for this anomaly, and given that the distance between sciences and non-sciences in Table 20.1 is much greater than the intra-scientific apparent anomalies, making the demarcatory qualitative difference clear.

From all this, it is clear that my method has a margin of error. This margin's value is very hard to estimate. Let us suppose for the sake of argument that my scores are only accurate to ± 10 points, which is a reasonable amount of inaccuracy (i.e., margin of error) given that mine is quite a crude method. Let us also suppose that in my results all this inaccuracy had systematically always fully favoured the sciences and disadvantaged the non-sciences, so that every non-science should have been valued 10 points higher than in Table 20.1, and every science should have been valued 10 points lower. Even in that case, I would still obtain results featuring the first two of the three levels of success described above, with every science analysed obtaining a positive score and every non-science analysed obtaining a score of 0 or less. This suggests that my approach is robust and that my method really does track scientificity and non-scientificity, and recovers the existing consensual extensional demarcation of the term "science". The best explanation for these results is that the compiled criteria used here really do demarcate science, and that my demarcatory approach works.

Hirvonen and Karisto (2022) have pointed out that, strictly speaking, we cannot know with certainty that the possession of a given set of properties is sufficient to establish scientificity in the way performed in this work. This is because my kind of analysis is defeasible in that it is non-monotonic, and therefore we have to always consider the possibility that the possession of an unknown or unconsidered property or set of properties (one not included in our Chap. 2) by a unit of analysis could reveal that that unit of analysis was/is in fact non-scientific. This caveat is technically correct, but such a revelation has not happened in any of the cases covered in my case studies presented here, and it seems implausible that it could happen for more than a tiny minority of possible units of analysis. And even if it were true that we do not or cannot know all possible properties relevant to a demarcation of science from non-science, this work's results suggest that in Chap. 2 we would still have a set of properties that is, in the vast majority of possible cases, sufficiently complete and representative to successfully execute such discernment.

We can therefore provisionally take the "cut" value for scientificity to be a value between−9 and 12. More case studies are necessary to be more accurate on this point, before setting a standard. Further research could lead us to be more specific about the "cut" value within the −9 to 12 range, or it might lead us to revise this range. This is because the clear cases of SBSs used in this analysis would have been more highly valued in Table 20.1 if their possession of properties of science were not taken for granted in the literature, and therefore 12 might not be the objectively correct minimum for scientificity. This implies that the gap between scientificity and non-scientificity might be even wider than evidenced in this work. There might also be a higher-scoring clear case of a non-scientific unit of analysis. It is clear than the contrast between the results of science and those of pseudoscience would be even starker if the properties of scientificity were not taken for granted in the literature when describing clearly scientific fields.

This work is a basic starting point of this analysis. Some authors may incorporate degrees, or argue that particular scores are incorrect for different reasons, or propose more precise scoring methods, or differential computing of different properties, or of the same property for different domains (if, for example, a given property were more important for some fields than for others) or for sciences in different stages of their respective developments. All that remains open. Any further work to refine the model or the method presented here to make them more accurate is worth undertaking, given that this basic approach, broadly speaking, has worked. In particular, any subsequent amendments to the −9 to 12 range might be arrived at through analyses of new units of analysis, or through refinements or extensions of the analyses already performed here. As above-mentioned, a work of the extension of this entire work could be written on each of the 17 individual cases of fields analysed.

The possible existence of a continuous spectrum going from non-scientificity to scientificity would not be an obstacle to setting a minimum standard for legitimately calling something scientific, not even for recognising science as a natural kind (cf. Mahner, 2007, pp. 521–22; 2013, pp. 39–40).

Different fields can be analysed with reference to the types of scientificity properties that they mostly possess or lack, in the light of the case studies just conducted. The following table (Table 20.2) summarises the results of our study regarding types of scientificity properties differentially possessed by, or lacking in, different non-scientific fields (the scientific CFs are not included in Table 20.2 because the fact that most properties of scientificity are taken for granted in the literature when describing clearly scientific fields renders their analysis in this respect unfruitful):

As one can observe in Table 20.2, the literature does not attribute overall epistemically-good practices to any of these non-scientific fields, with the exception of common sense. All of them, again with the exception of common sense, lack the epistemic and the formal properties found in scientific propositions, arguments, practices, and in the evolution of scientific CFs. The literature finds the contents and the epistemic attitudes and values in creation "science", homeopathy, and feng shui particularly faulty.

Following Thagard (1988), I have advanced a list of properties of science that belong to the "conceptual profile of science"; and realising his desideratum, I have provided "a list of features that are typical of science" (ibid., p. 159). In the process, I have also solved Kuhn's problem.

Table 20.2 Types of scientificity properties differentially possessed by, or lacking in, different non-scientific fields

	Astrology	Parapsychology	Creation "science"	Homeopathy	Feng shui	Common sense	Art	Religion
Properties related to practices, methods, and techniques	15 of 117	17 of 117	3 of 117	–	1 of 117	27 of 117	11 of 117	1 of 117
Formal properties of propositions, hypotheses, theories, and arguments	2 of 6	3 of 6	2 of 6	–	–	4 of 6	–	–
Properties related to the contents of problems, propositions, hypotheses, and theories	20 of 50	16 of 50	6 of 50	1 of 50	2 of 50	31 of 50	4 of 50	16 of 50
Properties related to epistemic attitudes and values	9 of 33	11 of 33	1 of 33	–	–	16 of 33	3 of 33	–

(continued)

Table 20.2 (continued)

	Astrology	Parapsychology	Creation “science”	Homeopathy	Feng shui	Common sense	Art	Religion
Epistemic properties of practices, propositions, hypotheses, and theories, and of the evolution of the discipline	1 of 14	3 of 14	–	–	–	4 of 14	–	1 of 14
Total of properties possessed	47	50	12	1	3	82	18	18

In addition, I have completed the Mahnerian project, i.e., (a) the composition of a broad "list" (set) of accepted characteristics, conditions, or properties of science, or indicators of scientificity (most of them, by themselves, unnecessary and insufficient) which may be collectively used to establish a demarcation between those theories, CFs, practices, etc. which are scientific and those which are not; and (b) the calibration of the list of properties, and, in particular, the ascertaining of a demarcatory boundary or "cut". All this was done by performing comparative analyses of a wide range of several clear cases of particular sciences and of non-sciences of different subtypes (natural sciences, social sciences, pseudosciences, non-pseudoscientific non-sciences, etc.).

In line with Boudry's "symptomatic approach" to the identification of pseudoscience (2019), my demarcation of science from all non-science is also "symptomatic", in that it takes indirect (and individually unnecessary) indicators of lack of epistemic reliability. In my account, what he calls "symptoms" and most of what he calls "immunizing strategies" (also in Boudry, 2011, 2013; Boudry & Buekens, 2011; Boudry & Braeckman, 2011, 2012; Boudry & Leuridan, 2011; Blancke et al., 2018, 2019) is lack of properties of science.

This accomplishment is a new one for the philosophy of science. If my procedure correctly demarcates every other clearly scientific and every other clearly non-scientific unit of analysis (i.e., all other CFs, as well as all theories, practices, attitudes, etc.), then the list of properties can appropriately specify and complete the intensional definition of SBS provided in Chap. 1: i.e., that science is that portion of intelligent activity (and its epistemic fruits) that is able to regularly lead an epistemic agent or community to hold beliefs that are sufficiently, and contextually the most, reliable, or at least the most intersubjectively epistemically justified (and better epistemically justified), on issues that are difficult to accurately evaluate (cf. Hansson, 2013, p. 70). It is my contention that it does.

Because not all non-science is pseudoscience, we still need to specify the properties that make some non-science pseudoscience. Section 1.1 has already defined "pseudoscience" as "a non-scientific doctrine or field that is presented by its theoreticians or practitioners as if it were scientific or a science, or that is easily taken to be so by a given epistemic community". There, I indicated that my intended intensional demarcation of pseudoscientificity is not exhausted by this definition, in part because it depends on the intensional demarcation of scientificity. But another part of the definition of pseudoscientificity is that, when the doctrine or field in question is presented by its theoreticians or practitioners as if it were scientific or a science, there is often involved a doctrinal component, immunising strategies (Boudry, 2011, 2013, 2019, 2022; Boudry & Buekens, 2011; Boudry & Braeckman, 2011, 2012; Boudry & Leuridan, 2011; Blancke et al., 2018, 2019), or social dynamics that are typical of communities of faith (see Koertge, 2013).

Because I have (apparently) found out a set of (known) properties of scientificity that makes all scientific CFs scientific in the same sense, i) it is possible to compare very different fields with regard to their respective degrees of scientificity; and ii) a single total ordering of, say, CFs with regard to their respective degrees of scientificity is possible. i) and ii) are possible even if these fields are scientific (or non-scientific) in many different ways.

The demarcation proposal presented here (that a given unit of analysis is scientific if and only if it possesses enough of the specific properties of science, and does not lack too many of the non-specific properties of science, in a manner such that the unit of analysis obtains a passing score under the scoring method specified in Chap. 2) solves all the problems that plagued all previous proposals, and its successful testing has been conducted against a wide sample of cases (fields). It is a sufficiently good solution as far as we currently know, and the best one we have. Therefore, it should be accepted, at least until a clear counterexample is presented. A clear counterexample would be a case of a unit of analysis (CF, method, practice, attitude, etc.) whose scientific status is agreed-upon, but whose status according to this demarcation method is as non-scientific; or vice versa. But even if a counterexample is found for one type of unit of analysis (e.g., practices), this demarcation proposal could still hold for other types of units of analysis (e.g., CFs, which is the type that has been studied in this work).

A consensus exists among the vast majority of philosophers and scientists about the extension of the term "science" (this, in turn, is strong evidence for believing that the notion of "science" can be explicated as a single concept). This work has provided us with an explication of the intension that is behind that existing extensional identification of clear cases of scientificity and of non-scientificity (together with a justification of that identification). This work has made explicit the complex intension that corresponds to the extension of the term "science" and that gives rise to said consensus. This explicitation has, in turn, given us an intensional epistemic demarcation of science: one that corresponds to, rationally reconstructs, explains, and explicates our existing philosophical and scientific consensus regarding the extension of the term "science". This demarcation is an EWSD. It is my contention that this work has provided us with a correct intensional demarcation of science, i.e., an intensional epistemic demarcation of science that captures the actual extension of the term "science", and nothing else, in a non-arbitrary manner. In any case, and pending further case studies, I have apparently arrived at an intensional epistemic demarcation of science that works as intended, for the first time in the history of the philosophy of science, therefore finally solving the main problem of SD. This result can hold i) independently of the truth value of my hypothesis that an intension of the term "science" already exists among the international philosophical and scientific community, and ii) independently of acceptance of the higher-level definition of the term "science" proposed in Sect. 1.4. It is surprising that no author before has attempted a survey like the one carried out in this work. Its results show that there is a clear ready-to-use demarcation of science, something that too many authors are sceptical about.

We can epistemically demarcate SBS from other intellectual activities and outputs. The philosophy of science should not be *merely* interested in studying what is *merely* deemed to be science; instead, it should strive to study science. Although what is science is clear in many cases, sometimes we would benefit from an analytical tool to discern it. This demarcation method is such a tool. If up until now our consensus on the extension of the term "science" seemed a bit arbitrary, unjustified, or uncritical, now it can be explicitly and analytically justified by using this demarcatory method, which is based on criteria that were already behind said extensional consensus. This allows all the 220 different and mutually-consistent insights from dozens and dozens of philosophers of science concerned with demarcation to be integrated and applied together. If successful, the demarcation developed here will be found to be empirically adequate in at least two manners, namely: 1) it will not give us false negatives or false positives on what is scientific with reference to the general consensuses in the literature,[2] and 2) its assessments will be empirically reliable (i.e., no contextually unreliable units of analysis will be labelled "scientific"). We would then be able to apply its analysis to contested, unclear, and new cases, i.e., we will be able to further build on the current reliable philosophical and scientific consensus, by going beyond our unaided ready recognition, and analytically demarcating with confidence in hitherto unclear cases. The analytically-developed tool provided in this work will then confirm its own worth by informing more reliable philosophical and scientific decisions, and judgments pertaining to a wide array of everyday human activities.

Were these results to keep being corroborated, we would be equipped with an explicit intensional SD that works as intended, for the first time in the history of the philosophy of science.

Science is often praisingly mentioned in educational, financial, medical, legal, regulatory, and political contexts. Courts of law and health care providers might give a special role to scientific evidence, and educational institutions might give a special role to science education. But this is problematic as things stand, given that there is no agreed-upon explicit intensional demarcation of science. Instead, we merely rely on whether the activity, product, treatment, piece of information, etc. is taken to be scientific or not by (a usually implicit or unstated) consensus.

The way of demarcating science proposed here constitutes a directly applicable practical decision procedure to assess the scientificity or non-scientificity of theories, CFs, practices, attitudes, testimonies, etc., able in principle to be utilised and applied in any context, for example, by governments, regulatory agencies, courts, tribunals, and health care and education providers.

Different degrees of stringency could be the most appropriate in different situations, or for the achievement of different goals. This SD procedure could be adapted to yield conclusions assigning non-binary degrees of scientificity to assessed units, and therefore different institutions could decide to require different degrees of scientificity in order to greenlight or to fund the same project, product, treatment, etc. Also, standards of scientificity featuring different degrees of stringency could be stipulated for purposes or matters of correspondingly different degrees of importance (e.g., a government could stipulate a requirement of a higher degree of

[2] This is: it will capture only and all units of analysis that fall under our philosophically-informed notion of "science", including its broadest normative criteria.

scientificity in matters related to human health than in matters only related to non-human animal health).

We need no longer labour under the idea that there is no working demarcation of science.

References

Blancke, S., Boudry, M., & Braeckman, J. (2018). Whence pseudoscience? An epidemiological approach. *Metode*. https://doi.org/10.7203/metode.8.10007

Blancke, S., Boudry, M., & Braeckman, J. (2019). Reasonable irrationality: The role of reasons in the diffusion of pseudoscience. *Journal of Cognition and Culture, 19*(5).

Boudry, M. (2011). *Here be dragons: Exploring the hinterland of science*. Ghent, Belgium: Ghent University. Faculty of Arts and Philosophy. https://biblio.ugent.be/publication/1191286

Boudry, M. (2013). The hypothesis that saves the day. Ad hoc reasoning in pseudoscience. *Logique et Analyse, 223*, 245–258.

Boudry, M. (2019). *Diagnosing pseudoscience in real life (a tribute to Sven Ove Hansson)*. [Preprint]. http://philsci-archive.pitt.edu/18630/

Boudry, M. (2022). Diagnosing pseudoscience – by getting rid of the demarcation problem. *Journal for General Philosophy of Science, 53*(2), 83–101.

Boudry, M., & Braeckman, J. (2011). Immunizing strategies & epistemic defense mechanisms. *Philosophia, 39*(1), 145–161.

Boudry, M., & Braeckman, J. (2012). How convenient! The epistemic rationale of selfvalidating belief systems. *Philosophical Psychology, 25*(3), 341–364.

Boudry, M., & Buekens, F. (2011). The epistemic predicament of a pseudoscience: Social constructivism confronts Freudian psychoanalysis. *Theoria, 77*(2), 159–179.

Boudry, M., & Leuridan, B. (2011). Where the design argument goes wrong: Auxiliary assumptions and unification. *Philosophy of Science, 78*(4), 558–578.

Fernandez-Beanato, D. (2020b). The multicriterial approach to the problem of demarcation. *Journal for General Philosophy of Science, 51*(3), 375–390. Erratum: Fernandez-Beanato, D. (2020c), "Correction to: The multicriterial approach to the problem of demarcation", *Journal for General Philosophy of Science, 51*(3), 391.

Fernandez-Beanato, D. (2021). Feng Shui and the demarcation project. *Science & Education, 30*(6), 1333–1351.

Hansson, S. O. (2013). Defining pseudoscience and science. In M. Pigliucci & M. Boudry (Eds.), *Philosophy of pseudoscience: Reconsidering the demarcation problem* (pp. 61–78). The University of Chicago Press.

Hirvonen, I., & Karisto, J. (2022). Demarcation without dogmas. *Theoria, 88*(3), 701–720.

Koertge, N. (2013). Belief buddies versus critical communities: The social organization of pseudoscience. In M. Pigliucci & M. Boudry (Eds.), *Philosophy of pseudoscience: Reconsidering the demarcation problem* (pp. 165–180). The University of Chicago Press.

Kuhn, T. S. (2000). *The road since structure: Philosophical essays 1970-1993*. University of Chicago Press.

Mahner, M. (2007). Demarcating science from non-science. In T. Kuipers (Ed.), *Handbook of the philosophy of science: General philosophy of science – Focal issues* (pp. 515–575). Elsevier.

Mahner, M. (2013). Science and pseudoscience. How to demarcate after the (alleged) demise of the demarcation problem. In M. Pigliucci & M. Boudry (Eds.), *Philosophy of pseudoscience: Reconsidering the demarcation problem* (pp. 29–44). The University of Chicago Press.

Thagard, P. (1988). *Computational philosophy of science*. MIT Press.

Appendix A: Fuzzy Logic for the Demarcation of Science

A.1 Introduction

As pointed out by Pigliucci (2013, p. 25), fuzzy logics are a possible tool to develop a multicriterial intensional demarcation of science from non-science, such as the multicriterial intensional epistemic demarcation proposed by Mahner (2007, pp. 521–2; 2013, pp. 29–43) and by me in this book.

Section 1.3.4 indicated that in many cases, saying, for example, that a unit of analysis (e.g., cognitive field, theory, research programme, practice, attitude, testimony, etc.) is "60 per cent scientific and 40 per cent non-scientific, and therefore both scientific and non-scientific at the same time and in the same sense" would be "more difficult to manage" but more accurate than providing a simple binary scientific demarcation (which would accord with classical logic and classify every unit assessed as simply scientific or non-scientific tout court).

The sense that we should make of a description of the type "60% scientific and 40% non-scientific" is closely analogic to the sense we would make of the description of a substance as "a mixed substance composed of 60% pure sand and 40% pure toothpaste": we might think that said substance does not qualify as "sand" tout court. We might think that it should not be used in mixtures for construction materials in which sand would usually be appropriately used. And we might think that it could be legitimately called "sand" and used as such for certain particular purposes if only it were 99.8% pure sand, instead of only 60%.

This first appendix proposes a system of fuzzy logic designed for said "more difficult" percentage management.

"[A] natural way to construct a fuzzy logic is as a many-valued logic with a continuum of truth values" (Priest, 2008, p. 224). In the most usual presentation of this type of construction of a standard fuzzy logic, the truth values, V, are the infinite set of the real numbers between and including 0 and 1—that is to say that for any valuation v, $0 \leq v \leq 1$—, where 1 means "completely true", and 0 means "completely

D. Fernandez-Beanato, *A Working Scientific Demarcation*, Synthese Library 519, https://doi.org/10.1007/978-3-031-99005-2

false", 0.5 means "true only at 50%", etc. (ibid.). This would allow us to capture degrees of scientificity with high precision, by contrasting units of analysis against accepted properties of science or of scientificity (as done in this book). The valuation in fuzzy logics of the truth value of propositions of the form Sa—that is, in our semantics, propositions which indicated that a certain unit of analysis *a* is scientific—would directly and intuitively reflect the degree of scientificity that we attribute to that *a*: 1 will be the valuation corresponding to the complete satisfaction of each and every one of our accepted, applicable criteria of scientificity, while 0 will be the valuation corresponding to the complete dissatisfaction of all of them.

Classical logics would lead us to classify each unit of analysis as simply scientific or non-scientific tout court, as the semantics of a bivalued logical system (which classical logics are) would require. In bivalued logical systems, the only valuations that would be assignable to a proposition of the form Sa would be "(completely) true", which, interpreted, would mean "*a* is (completely) scientific"; and "(completely) false", which, interpreted, would mean "*a* is (completely) non-scientific".

But when executing a multicriterial scientific demarcation, we might want to assign some units of analysis valuations v of the type $0 < v < 1$ (i.e., greater than 0 and less than 1). In this type of valuation, the complement of a degree of scientificity of less than 1 will be a certain degree of non-scientificity greater than 0 (for example, a cognitive field d could be valuated as scientific in grade 0.1 only, which would indicate that d is scientific at 10% and non-scientific at 90%).

Fuzzy logical systems are more powerful than classical ones in the sense that the former are not only able to express the valuations v 0 and 1, but also intermediate valuations (i.e., the ones of the type $0 < v < 1$).

An understanding of many different lower and higher degrees of scientificity, i.e., of values v, should be developed via analytico-empirical investigation, along the lines of what is presented in this book, of the cumuli of properties featured in clear cases of agreed-upon, mature, successful scientific units of analysis, and different degrees of divergence from those cumuli.

In case it was advisable, we could then translate intermediate valuations into valuations of either truthfulness simpliciter or falsehood simpliciter (or of either acceptability or inacceptability) assigned to the propositions of the form Sa.

A.2 Proposal, Use, and Examples

In order to do this, we need another system, one for the introduction of discrete borders into the value continuum. My proposal for such a system is to define the acceptability of a statement of the form Sa as having a "cut" truth-value equal or greater than c where $0 < c < 1$ (a truth value of 0 would, of course, valuate a unit of analysis as clearly and totally non-scientific, while a truth value of 1 would valuate it as clearly and maximally scientific). The "cut" value c, in turn, should be ascertained by performing comparative analyses of a wide range of several clear cases of particular sciences and of non-sciences of different subtypes (natural sciences, social sciences, pseudosciences, non-pseudoscientific non-sciences, etc.). This process should yield a (preferably universal) semantic consensus (of the form "This is the 'cut' value at which a unit of analysis of type *t* should start to be accepted as scientific in society *s*"). Such consensus could be arrived at because there is an almost complete lack of dispute among authors with regard to most of the properties of science (see Sect. 2.1, also Sect. 1.3.4).

Among the operations that are admissible in fuzzy logics are negation, conjunction, and disjunction. Their valuations, according to the proposal of Zadeh (1965), are as follows:

$$V(\neg A) = 1 - V(A)$$

(Therefore, the following three expressions are tautologies:

$$V(A) + V(\neg A) = 1$$

$$[V(A) = 1] \Rightarrow [V(\neg A) = 0]$$

$$[V(A) = 0] \Rightarrow [V(\neg A) = 1])$$

$$V(A \wedge B) = \min(V(A), V(B))$$

$$V(A \vee B) = \max(V(A), V(B))$$

But, for our purposes, we must modify Zadeh's proposal regarding disjunction, in the following way:

$$V(A \vee B) = \max(V(A), V(B)) \textbf{ except } \text{if A is the negation of B,}$$
$$\text{in which case } V(A \vee B) = 1$$

This last modification is because, as in classical logics, here, for any given *a*, "S*a* ∨ ¬S*a*" is tautological, given that the principle of the excluded third applies. This is, in turn, due to the fact that every possible unit of analysis will be either scientific or non-scientific, or a mixture of these two possibilities, without anything else being possible. "Grey" or borderline cases or areas would, then, be simply cases of epistemic opacity or ontic mixtures of the scientific and the non-scientific, not a third option.

The usefulness of the application of these operations to multicriterial scientific demarcation is undeniable. Negation would allow us to establish the respective complements of scientificity and of non-scientificity of a given unit of analysis. Conjunction would allow us to establish the degree of scientificity of any family of disciplines, theories, cognitive fields, scientific research programmes, etc. that had not been valuated directly as an aggregate. For example, if Lacanian psychoanalysis had a degree of scientificity of 0.2, and Freudian psychoanalysis had a degree of scientificity of 0.1, and if the conjunction of these two subdisciplines constituted the entirety of the discipline of psychoanalysis, then one could establish the degree of scientificity of psychoanalysis as a whole, because in fuzzy logics the value of a composite proposition can be determined by the value of its component propositions (cf. Cintula et al., 2016). Thus, in this case, the value assigned to the proposition "Lacanian psychoanalysis is scientific and Freudian psychoanalysis is scientific" will be equal to that assigned to the proposition "Psychoanalysis is scientific" (in our example, 0.1).[1] Lastly, disjunction would allow us to establish whether or not there is at least a sufficiently scientific part in a set of interrelated disciplines, theories, or cognitive fields of low aggregated degree of scientificity. Of course, as explained above, the value assigned to a tautological proposition such as "Astrology is scientific or astrology is not scientific" will always be 1.

If a unit of analysis is slightly more scientific than non-scientific, that might not necessarily mean that said unit is scientific tout court. As an illustration of this, consider that we might disagree with the contention that it is legitimate to use the term "sand" to refer to a mixed substance composed of 51% pure sand and 49% pure toothpaste. Architects and engineers using sand as a construction material in important projects certainly would.

A criticism that can be made of the use of a fuzzy logical system for scientific demarcation is that it allegedly adds an unnecessary complication to the resolution of the problems of scientific demarcation, by proposing to use a logic which is not classical. Mahner seems to implicate this criticism by pointing out that classical logics would suffice to execute a multicriterial scientific demarcation (Mahner, 2013, p. 42). However, using non-classical logics is an in-principle problem only for pro-classical logical monists. And it is not necessary for a scientific-demarcationist to accept only classical logics as correct. Therefore, the use of a non-classical logic is not necessarily a problem, neither in principle nor in practice.

That of scientific demarcation is a project that can be equipped, without loss, with the logical system that allows it to be executed in the best possible way, without having a particular need to remain tied to classical logics. Such a need could not be justified solely on the grounds that classical logics are the most universally accepted or the most used type of logics in most fields of formal application of logics. One could even imagine a plausible situation in which the success of a scientific demarcation executed by using fuzzy logics would raise the level of general acceptance of fuzzy logics.

[1] If we tried to do this with a classical logic using the system proposed by Mahner (2013, pp. 29–43), the imprecision contained in the valuations of the particular subdisciplines (due to the bivaluation of classical logics) would be transmitted to the valuations of the aggregated disciplines.

Appendix B: Social Beliefs Without Entirely Social Causes

If the Strong Programme in the Sociology of Scientific Knowledge's positions are considered for the philosophical realm (instead of just the sociological realm), other problems emerge that make it inadequate for our demarcatory purposes, given that the demarcation looked after here is one that, hopefully, holds across cultures and time periods. This second appendix describes some of these problems.

The Strong Programme's requirement of symmetry establishes that the same types of causes should mainly explain all (both true and false) beliefs. This appendix argues that the main problem with this requirement is that, because it is not restricted to social causes, it violates the empiricity of sociology, as it is aprioristic.

The defenders of the Strong Programme highlighted the need and relevance of having a sociological study of scientific beliefs. They also indicated that the sociology of scientific knowledge should causally explain the adhesion of a particular society to the particular scientific beliefs held by that society.[1] This is not heterodox at all: as Kukla (2000, p. 10) has pointed out, the programme of studying the social causes of scientific beliefs does not really have a substantial opposition, unless straw men images are built to present the Strong Programme under an excessively novel light.

Barry Barnes (1943–), a leading promoter of the Strong Programme, argued in *Interest and the Growth of Knowledge* that a social study of scientific beliefs is essential because the set of beliefs held by a given society depends on the practices that that society executes and the interests moving it, as well as on the resources

[1] They understood "causal explanation" of belief as an explanation "concerned with the conditions which bring about belief" (Bloor, 1991, p. 7). This last desideratum is born from a causal-explanatory scientific methodological monism (cf. Tozzi, 2003, p. 116): in *Knowledge and Social Imagery*, Bloor argued that the detection of causal relationships is one of the "values which are taken for granted in other scientific disciplines", so a sociology of knowledge must be able, as a science, to provide such explanations in order to position itself in a situation of parity with these other scientific disciplines (Bloor, 1991, p. 7).

D. Fernandez-Beanato, *A Working Scientific Demarcation*, Synthese Library 519, https://doi.org/10.1007/978-3-031-99005-2

(material, conceptual, etc.) available to it (Barnes, 1977, pp. 1–2). Barnes and David Bloor (1942–), another referent of the Strong Programme, pointed out that the social use of these resources in the development of scientific beliefs makes it clear how inappropriate it can be to conceive of rationality as a property of individuals considered in isolation, and the erroneousness of proposing a dichotomy between the rational and the social (cf. Tozzi, 2001, p. 529). In the first chapter of *Interest and the Growth of Knowledge*, Barnes also presents the following positions as distinctive of his conception and as alien to the "progressive realist" conception, which he calls "orthodox" and "positivist" (and "rationalist" in Barnes & Bloor, 1982, p. 21): that science is the product of a historical development, that it suffers contextual limitations related to the cognitive cultural resources available to the community that practices it, that it depends on cultural backgrounds or cultural antecedents, that in it the factual is often intertwined with the theoretical, and that scientific activity comprises pragmatic, applicative, technical, technological, and experimental aspects. But, again, few current philosophers, including "orthodox progressive realists", would deny any of these propositions.

The founders of the Strong Programme concluded that they defended a "relativist theory of knowledge" (Barnes & Bloor, 1982, p. 21), while maintaining a non-factive concept of "knowledge".

Furthermore, the Strong Programme defends ontological realism: Barnes recognizes that "[e]verything of naturalistic significance would indicate that there is indeed one world, one reality, 'out there', the source of our perceptions if not their total determinant, the cause of our expectations being fulfilled or disappointed, of our endeavours succeeding or being frustrated", and that this world is "the source of *primitive causes*" that produce changes in our beliefs (Barnes, 1977, p. 25, emphasis in the original). Bloor wrote, as part of the Strong Programme's first tenet: "Naturally there will be other types of causes apart from social ones which will cooperate in bringing about belief" (Bloor, 1991, p. 7; cf. Barnes, 1977, p. 25). But almost no philosophers rule out a role of social causes in these processes either.

Most philosophers can in principle agree with two basic starting points of the Strong Programme, namely: (i) "the observation that beliefs on a certain topic vary", and that ii) "which of these beliefs is found in a context given depends on, or is relative to, the circumstances of the users" (Barnes & Bloor, 1982, p. 22). Most will not necessarily have problems to admit three of the four methodological principles proposed by the creators of the Strong Programme (cf. Bloor, 1991, p. 7), namely: i) that sociology should seek causal explanations; ii) that in the case of opposite phenomena, such as rationality and irrationality, or success and failure, both require explanations; and (skipping the third one) iv) that sociology, as a social science, must be reflexive, that is, it must apply to itself the general patterns of explanation that it uses.

The remaining programmatic tenet of the Strong Programme, iii), is the so-called requirement (or principle) of symmetry (or equivalence). This is the key element that makes this Programme unique. It establishes that the sociology of knowledge must be symmetric in its "style" of explanation, that is, it must explain all beliefs, whether true or false, by the same types of (main) causes (Bloor, 1991, p. 7). This third programmatic principle summarizes the kind of social "relativism" that they promoted (Barnes & Bloor, 1982, pp. 22–3). Barnes and Bloor also believe that this symmetry necessarily implies a commitment to epistemic relativism (the thesis that there are no absolute, trans-social, and trans-historic rational justifications to support certain beliefs). Kukla (2000, pp. 9–18) convincingly argues that this is not the case.

The Strong Programme's principle of symmetry is inconsistent with anti-social-relativism in sociology of science only if it implies that social causes must be the *main* explanation of all types of beliefs. In this regard, the defenders of the Strong Programme have been ambiguous. Sometimes, it seems that the principle of symmetry is limited to establishing that the same types of *social* causes must explain all types of beliefs *as long as we are speaking only of those causes that are social.* The context of Bloor's explicit formulation of the principle of symmetry (Bloor, 1991, p. 7) can seem to favour this interpretation, as does Barnes when he emphasizes:

> What matters is that we recognise the *sociological* equivalence of different knowledge claims. We will doubtless continue to evaluate beliefs differentially ourselves, but such evaluations must be recognised as having no relevance to the task of sociological explanation; as a methodological principle we must not allow our evaluation of beliefs to determine what form of sociological account we put forward to explain them. (Barnes, 1977, p. 25, emphasis in the original)

Anti-social-relativists in the sociology of science can readily admit that the sociology of science is not restricted to being just a sociology of error, and that, instead, some realities of the world are *non-social* causes in the formation of true or truthlike beliefs *in appropriate social contexts*, while in other social contexts this does not happen so easily. They could perfectly, therefore, agree with the principle of symmetry if it were restricted to social causes. For example, anti-social-relativists in the sociology of science could accept that false beliefs held in a social configuration that favours the formation and maintenance of false beliefs can receive the same type of *sociological*[2] causal explanations that true or truthlike beliefs held in a social configuration that favours the formation and maintenance of true or truthlike beliefs.

But the interpretation given by the Strong Programme's main authors to their principle of symmetry is not restricted to social causes; instead, it establishes that the same types of causes *in general* (not only the same types of *social* causes) should mainly explain all types of beliefs. In other words, that "all beliefs are on a par with one another with respect to the causes of their credibility [...] regardless of truth and falsity" (Barnes & Bloor, 1982, p. 23). They indicate:

> For the relativist there is no sense attached to the idea that some standards or beliefs are really rational as distinct from merely locally accepted as such. Because he thinks that there are no context-free or super-cultural norms of rationality he does not see rationally and irrationally held beliefs as making up two distinct and qualitatively different classes of things. They do not fall into two different natural kinds which make different sorts of appeal to the human mind, or stand in a different relationship to reality, or depend for their credibility on different patterns of social organization (Barnes & Bloor, 1982, pp. 27–8)

---

[2] That is, with respect to the exclusively *social* causes of their credibility.

So, for sociologists adhering to the Strong Programme, it would not be correct to interpret the letter of the principle of symmetry as indicating that a) there are *social* causes for all beliefs, and that b) all those causes must be analysed within sociology in the same way, but that c) when it comes to causes in general, *social* causes may not be the main ones to adequately explain the emergence and maintenance of true or truthlike beliefs. Interestingly, according to this conception, even in the case of a society whose social configuration favoured the development of true or truthlike beliefs, the reality of the world outside of that society would not be a main cause of the appearance of true or truthlike beliefs in that society. That is why, for the Strong Programme, the mere existence of a world external to societies does not help philosophers to find trans-social or trans-historic canons of rationality.

Kukla is correct again in pointing out that Barnes and Bloor rushed to equate their kind of social "relativism", that of the principle of symmetry, to epistemic relativism (and hence their "strong" or "generalist" interpretation of that principle), without having provided a justification for said conceptual operation (Kukla, 2000, p. 14). It is due to the latter that our analysis does not yield a complete dissolution of the theoretical disagreements between the Strong Programme and anti-social-relativists in the sociology of science.

In summary, the original literal version of the symmetry principle of the Strong Programme can be accepted even by anti-social-relativists in the sociology of science. In principle, none of them is forced to deny that the formation and maintenance of both true and false social beliefs should receive the same types of *sociological* explanations (that is, explanations in terms of purely social causes), provided that it is admitted that this type of causes should not be deemed a priori, that is, in all cases, hegemonic over other possible types of causes. The problem arises when the theorists of the Strong Programme take a seemingly unjustified leap from the proposal of the principle of symmetry to the affirmation of epistemic relativism, with which social explanations become a priori hegemonic in all cases. This proposal of an apriorism for the analysis of empirical problems is another serious weakness of the Strong Programme (cf. Kukla, 2000, pp. 8, 10). This leap was not required by the original formulations of the four main theses of the Strong Programme, and accepting it would neutralize what is legitimate and promising in any anti-social-relativist sociology of science.

References

Barnes, B., & Bloor, D. (1982). Relativism, rationalism and the sociology of knowledge. In M. Hollis & S. Lukes (Eds.), *Rationality and relativism* (pp. 21–47). MIT Press.

Barnes, B. (1977). *Interest and the growth of knowledge*. Routledge.

Bloor, D. (1991). *Knowledge and social imagery* (2nd ed.). University of Chicago Press.

Cintula, P., Fermüller, C. G., & Noguera, C. (2016). Fuzzy logic. In E. N. Zalta (Ed.), *The Stanford Encyclopedia of Philosophy* (Winter 2016 ed.). Stanford University. https://plato.stanford.edu/archives/win2016/entries/logic-fuzzy/

Kukla, A. (2000). *Social constructivism and the philosophy of science*. Routledge.
Mahner, M. (2007). Demarcating science from non-science. In T. Kuipers (Ed.), *Handbook of the philosophy of science: General philosophy of science – Focal issues* (pp. 515–575). Elsevier.
Mahner, M. (2013). Science and pseudoscience. How to demarcate after the (alleged) demise of the demarcation problem. In M. Pigliucci & M. Boudry (Eds.), *Philosophy of pseudoscience: Reconsidering the demarcation problem* (pp. 29–44). The University of Chicago Press.
Pigliucci, M. (2013). The demarcation problem. A (belated) response to Laudan. In M. Pigliucci & M. Boudry (Eds.), *Philosophy of pseudoscience: Reconsidering the demarcation problem* (pp. 9–28). The University of Chicago Press.
Priest, G. (2008). *An introduction to non-classical logic: From if to is*. University of Cambridge Press.
Tozzi, M. V. (2001). Malos entendidos en torno al Programa Fuerte. *Epistemología e Historia de la Ciencia, 7*(7), 527–532.
Tozzi, M. V. (2003). Wittgenstein y la sociología del conocimiento. In S. Cabanchik, F. Penelas, & M. V. Tozzi (Eds.), *El giro pragmático en la filosofía contemporánea* (pp. 115–130). GEDISA.
Zadeh, L. A. (1965). Fuzzy sets. *Information and Control, 8*(3), 338–353.

Index

D. Fernandez-Beanato, *A Working Scientific Demarcation*, Synthese Library 519, https://doi.org/10.1007/978-3-031-99005-2

B

D

P

Q

R

MIX
Papier aus verantwortungsvollen Quellen
Paper from responsible sources
FSC® C105338

If you have any concerns about our products,
you can contact us on
ProductSafety@springernature.com

In case Publisher is established outside the EU,
the EU authorized representative is:
Springer Nature Customer Service Center GmbH
Europaplatz 3, 69115 Heidelberg, Germany

Printed by Libri Plureos GmbH
in Hamburg, Germany